A Greater Freedom

A Greater Freedom

Biotechnology, Love, and Human Destiny

In Dialogue with
Hans Jonas and Jürgen Habermas

BY

Stephan Kampowski

WITH A FOREWORD BY

Stanley Hauerwas

PICKWICK *Publications* · Eugene, Oregon

A GREATER FREEDOM
Biotechnology, Love, and Human Destiny (In Dialogue with Hans Jonas and
Jürgen Habermas)

Pickwick Publications
An Imprint of Wipf and Stock Publishers
199 W. 8th Ave., Suite 3
Eugene, OR 97401

www.wipfandstock.com

ISBN 13: 978-1-61097-900-9

Cataloging-in-Publication data:

Kampowski, Stephan.

A greater freedom : biotechnology, love, and human destiny (in dialogue with Hans Jonas and Jürgen Habermas) / Stephan Kampowski ; foreword by Stanley Hauerwas.

xviii + 186 p.; 23 cm—Includes bibliographical references.

ISBN 13: 978-1-61097-900-9

1. Biotechnology—Moral and ethical aspects. 2. Philosophical anthropology. I. Hauerwas, Stanley, 1940–. II. Title.

BL256 .K33 2013

Manufactured in the USA.

The author and publisher gratefully acknowledge permission to reprint material from the following sources:
Excerpts from Jürgen Habermas, *The Future of Human Nature* (Cambridge, UK: Polity Press 2003). Used by permission of Polity Press.
Excerpts from Hans Jonas, *The Phenomenon of Life: Toward a Philosophical Biology,* Northwestern University Studies in Phenomenology and Existential Philosophy (Evanston, Ill.: Northwestern University Press, 2001). Used by permission of Northwestern University Press.

Contents

Contents

Foreword

By Stanley Hauerwas

WE LIVE IN A time in which many seem determined to use intellectual power to get out of life alive. Ironically, the fear of death creates a culture of death, in which those who have means are able to impose their fear of death onto those who do not have the means to cope with those fears. As a result, what has been called the technological imperative now seems to have become a permanent feature of modern life: what can be done should be done.

We are, therefore, extremely fortunate to have this book by Stephan Kampowski on the work of Hans Jonas and Jürgen Habermas. By drawing on the work of Jonas and Habermas, Kampowski helps us see that the technological imperative cannot help but corrupt our humanity because, with its utopian hopes, it inspires us to live as if we were our own creators. As a result, we live lives of loneliness.

With extraordinary erudition and philosophical acumen, Kampowski makes the work of Jonas and Habermas available to a wide range of readers. Given that he is such a faithful expositor of Jonas' and Habermas' work, some might be tempted to consider this a book primarily about these thinkers. It is certainly a wonderful introduction to their philosophy. But to interpret this volume primarily as an introduction to the work of Jonas and Habermas would do injustice to its significance. What Kampowski helps us see is that a philosophical case can be made against the technological imperative. This is a voice we desperately need.

One of this work's important contributions is Kampowski's explanation of how Jonas' account of the teleological character of the organism is interrelated with the ethic of responsibility. The argument that life is a determinative category by which we are able to understand the character of all existence is crucial if we are to recover the limits necessary for shaping our care of one another through the agency of science. Kampowski provides a wonderful account of Jonas' understanding of death and how death rightly should shape our living in order to reveal our perverse accounts of freedom.

One of the virtues of this book is Kampowski's ability to bring these thinkers into conversation with one another, but also to bring other voices into the conversation, the voice of Robert Spaemann in particular. Spaemann is a philosopher who has done fundamental work from which we, particularly in the English speaking world, need to learn. Kampowski's breadth of knowledge of contemporary European philosophical traditions means he is able to use Spaemann to bring light to both the strengths and limits of Jonas and Habermas. His suggestion, drawing on the work of Gerald McKenny, that Jonas' account of responsibility itself betrays some utopian desires is extremely important. As a result, Kampowski helps us see that the voice of theology has much to contribute if we are to have an alternative to the culture of death.

This is a book that hopefully will be widely read by Christians and non-Christians alike. Kampowski's philosophical analysis of Jonas is one that no Christian can afford to ignore if we are to speak truthfully to the world in which we find ourselves.

Acknowledgments

THIS BOOK IS ESSENTIALLY the fruit of the three years that I was allowed to be a "professore a contratto" at the Pontifical John Paul II Institute in Rome, a position that entails a relatively light teaching load and provides the possibility of engaging in a post-doctoral research project.

The bulk of the work, however, was done in the Spring of 2008, which I spent as a visiting scholar at Duke University in Durham, North Carolina in the United States. There I had the great privilege of benefitting from the direction and advice of Dr. Stanley Hauerwas, for whose help I am extremely grateful.

My sincere appreciation also goes to Dr. L. Gregory Jones, the Dean of Duke Divinity School, for the kind invitation addressed to me and for making Duke's wonderful academic resources available to me, and to Dr. Reinhard Hütter and his family, who have given me great encouragement and support during my stay at Duke.

I should also like to express my profound gratitude to Mons. Livio Melina, the President of our Institute, for releasing me from my other academic responsibilities during the Spring 2008 semester so that I could take the opportunity of spending this time in North Carolina.

I am indebted to the whole academic community of the John Paul II Institute for the lively and inspiring exchange of ideas that has occurred throughout these years. In particular I would like to mention Dr. Stanisław Grygiel, the Director of the Institute's *Cattedra Karol Wojtyła* Research Chair, together with all those participating in the activities of the *Cattedra,* and Rev. José Noriega, the then-Vice-President of our Institute.

Finally, I would like to thank the students of my seminars on Hans Jonas (Fall 2006) and on Jürgen Habermas (Fall 2007) whose questions and insights helped my own deeper understanding of these authors.

Introduction

Freedom, Technology, and Destiny

"THE NEXT FRONTIER IS . . . ourselves."[1] The West has been won and the moon has been conquered. The human person's vigorous spirit needs a new task against which to measure itself. And it seems that Gregory Stock is right. The new frontier is not Planet Mars but something by far more challenging, promising, and fascinating: it is ourselves. In the past two hundred years or so, technology has accomplished incredible feats in transforming the world, and to many it seems that the time has come to apply our attempts at changing the world to ourselves: "Starting with fire and clothes, we looked for ways to ward off the elements. . . . Telephones and airplanes collapsed distance. Antibiotics kept death-dealing microbes at bay. Now, however, we have started a wholesale process of aiming our technologies inward. Now our technologies have started to merge with our minds, our memories, our metabolisms, our personalities, our progeny and perhaps our souls. Serious people have embarked on changing humans so much that they call it a new kind of engineering evolution—one that we direct for ourselves."[2]

At the same time, while technology remains fascinating and no one would want to miss its many accomplishments, we have generally grown more sober in its regard, having learned the hard way the thoroughly dialectical character of our technological ingenuity. Modern technology, essentially a child of the Enlightenment, has its share in the famous dialectics of the latter.[3] There is, first of all, the fact that yesterday's privileges become today's necessities, so that our new toys do not always make us freer and happier but rather create new dependencies, multiplying our

1. Stock, *Redesigning Humans*, 171.
2. Garreau, *Radical Evolution*, 6.
3. Cf. Horkheimer and Adorno, *Dialectic of Enlightenment*.

reasons to be unhappy. Just a few decades ago, who would have felt upset when, due to a temporarily failed Internet connection, it took half an hour to send an important document from Rome to New York instead of the usual two minutes? We easily take for granted the marvels of instant communication and get frustrated when for a brief period we are negated its benefits. Without necessarily having in mind the Internet, already Jean-Jacques Rousseau observed these dynamics: "For, besides their continuing thus to soften body and mind, as these commodities had lost almost all their pleasantness through habit, and as they had at the same time degenerated into true needs, being deprived of them became much more cruel than possessing them was sweet; and people were unhappy to lose them without being happy to possess them."[4] Technology, moreover, can be used for good or ill, and even in its essentially benign uses it often has bad side effects, which may well outweigh the benefits they bring with them.

All this goes to say that the promises of winning the new frontier, that is, of applying our technology to ourselves, fall onto soil that is generally more critical of technological progress than has been the case in the past, in a world prior to Chernobyl and global warming. And yet, the appeal of biotechnology seems hard to resist: "Rather than fearing change, we ought to embrace it, rather than prohibiting the exploration of new technologies, society ought to focus on spreading the power to alter our minds and bodies to as many people as possible. . . . The benefits to be won from biotechnology are concrete and measurable. Keeping people young longer would slow the rise in worldwide health spending . . . . Improving human memory, attention, and communication abilities would increase productivity, which in turn would lead to new scientific discoveries and faster innovation."[5]

Then again, when we listen to critics of the biotechnological revolution, we certainly find arguments one could in principle level against any new technology, such as appeal to cautionary principles and cost-benefit calculations. But we also find concerns so fundamental that no one in his or her right mind would ever raise them against the use of airplanes or the composing of short text messages. These fears give expression to the fact that here, with biotechnology applied to human beings and their very nature, we are indeed heading toward a new frontier, which raises issues

4. Rousseau, "Discourse on the Origin," 147.
5. Naam, *More Than Human*, 5–6.

of an unprecedented kind. Thus, Francis Fukuyama, in his best-selling *Our Posthuman Future*, voices the concern that what could be at stake here is our very humanity along with our moral sense: "The deepest fear that people express about technology is . . . that, in the end, biotechnology will cause us in some way to lose our humanity. . . . Human nature is what gives us a moral sense, provides us with the social skills to live in society, and serves as a ground for more sophisticated philosophical discussions of rights, justice, and morality. What is ultimately at stake with biotechnology is . . . the very grounding of the human moral sense, which has been a constant ever since there were human beings."[6]

In his booklet *The Future of Human Nature*, Jürgen Habermas echoes Fukuyama's concern, wondering what an established practice of biotechnological engineering would do to our moral self-understanding: "Will we still be able to come to a self-understanding as persons who are the undivided authors of their own lives, and approach others, without exception, as persons of equal birth? With this, two presuppositions of our moral self-understanding . . . are at stake."[7] In this way, both authors express an existential and dramatic concern: with our biotechnology we may risk abolishing the human person as a moral being.

How could this be so? Do these authors not perhaps overstate their case? What is at the basis of these fears? In this book, I will attempt to put the promise of biotechnology, which mainly consists in a greater freedom by giving us greater strength, superior intelligence, and more years to live, into the perspective of our human destiny, which, I will argue, consists in love. The greatest freedom is the freedom for our destiny, which is the freedom to love. But love is nothing that can be manufactured or technologically enhanced. It does not as such fall under the objects of biotechnology, while biotechnology, at least in some of its forms, may make it more difficult for us to love and hence, it may actually decrease our freedom.[8]

In order to explore this hypothesis, I will turn to the thought of Hans Jonas, one of the founding fathers of what today is called "bioethics."[9]

---

6. Fukuyama, *Our Posthuman Future*, 101–2.

7. Habermas, *Future of Human Nature*, 72.

8. Cf. the concern that Stanley Hauerwas raises in this context: "For when freedom and its enhancement becomes an end in itself, we lose any account of human life that gives content and direction to freedom. As a result we end by being less rather than more free" (Hauerwas, *Suffering Presence*, 14).

9. Jonas was one of the founding Fellows of the Hastings Center on Bioethics, the

In the first chapter, I will examine his "philosophy of the organism."[10] Jonas explains the distinguishing characteristics of the living body, which make manipulating an organism very different from manipulating life-less things. I hope to show with Jonas that the meaning of the organism is freedom understood as the power of self-transcendence. Freedom in turn—and here I will go beyond Jonas, while moving from his premises—finds its highest expression in love understood as a call to communion.

In the second chapter I will discuss Jonas' philosophy of responsibility, arguing that for him responsibility ultimately amounts to benevolence and that benevolence is at the foundation of the new categorical imperative he proposes: "Act so that the effects of your action are compatible with the permanence of genuine human life."[11] For Jonas humanity ought to be because responsibility ought to be. In other words, the reason for why it is better for humanity to be rather than not to be lies in this: only with humanity there is the principle of responsibility and benevolence in the world. A world in which there is benevolence or love is better than one in which these are absent. The greatest concern about biotechnology that Jonas voices, along with Fukuyama and Habermas who follow him here, is that one day, with our tools, we may prevent our descendants from being responsible or benevolent beings. Jonas writes, "[It is] their duty over which we have to watch, namely their duty to be truly human," which—as becomes clear in the rest of Jonas' book—amounts to their capacity to be responsible beings, a capacity of which we could rob them "with the alchemy of our 'utopian' technology."[12]

In the final chapter, I will turn to Jürgen Habermas and the way in which he engages Hans Jonas in his essay *The Future of Human Nature*, spelling out in more detail the dangers of injuring or even abolishing human morality by biotechnology. Some aspects of Habermas' argument presuppose peculiar elements of his theory of communicative action and discourse ethics, while other important parts draw on Jonas' thought. The main line of his reasoning consists in showing how, by means of biotechnology, one generation may attempt to impose its own ideas and intentions on the next generation. This imposition disrupts the equality that previously existed among them, dividing them into one master

first American "think tank" for bioethics (Cf. L. R. Kass, "Practicing Ethics," 5–12 and Jonas, *Memoirs*, 200).

10. Cf. Jonas, *Phenomenon of Life*.

11. Jonas, *Imperative of Responsibility*, 11.

12. Ibid., 42.

generation and many manufactured ones who are deprived of the full extent of their spontaneity and freedom and will have to feel inferior toward those who made them, no longer able to have a sense of full authorship of their lives. Summing up both Jonas' and Habermas' thought, one can say that the central issue consists in this: we must not impose our own image on our descendants. In fact, the Bible's prohibition against the making of an image (cf. Exod. 20:4) can with very good reason be applied not only to the Lord but also to the human person made in his image, which for us will always remain mysterious and out of the reach of our free disposal.[13]

The Context and Procedure of Our Study

Before going into the argument, let us briefly discuss the context of Hans Jonas' life and work. Jonas was born in Germany in 1903 and studied under luminaries such as Edmund Husserl, Martin Heidegger, and Rudolf Bultmann at the universities of Freiburg, Berlin, and Marburg.[14] As he presented a paper in one of Heidegger's seminars, the latter was so excited about it that he helped him to get it published. *Augustin und das paulinische Freiheitsproblem*—"Augustine and the Pauline Problem of Freedom"—would thus become Jonas' first book. For his dissertation, which he wrote under Heidegger's direction, he turned to a study of the Gnostic religion. His *Gnosis und spätantiker Geist* became an influential, if not uncontroversial, work on the subject.[15] When the National Socialists took power in Germany in the 1930's, Jonas, who was Jewish and active in the Zionist movement, first went to England for a year and then emigrated to what was then Palestine. At this time, he vowed to himself that only as a soldier of a conquering army would he ever set foot again in Germany.[16] In 1940 Jonas volunteered for the Jewish Brigade of the British Army, with which he in fact victoriously entered Germany in 1945. Already during the war, as he was separated from his books, his interest shifted from Gnosticism to what he would later call a "philosophical

13. I owe the idea of summing up their arguments by means of the above Scripture passage to Junker-Kenny, "Genetic Enhancement," 12.

14. For most of the biographical data, see his autobiography: Jonas, *Memoirs*.

15. Jonas, *Gnosis und spätantiker Geist*; English: *The Gnostic Religion*.
For an apt summary of the controversies ensuing upon the publication of Jonas' work and for an appraisal of its significance for Gnosis research, see Waldstein, "Hans Jonas' Construct 'Gnosticism,'" 341–72.

16. Jonas, *Memoirs*, 75.

biology." Perhaps it was the experience of the utter precariousness of life during the war that prompted him to turn his attention to issues related to the organism, i.e., to the question of what it means for something to be "alive."[17] In 1948 he took part in Israel's War of Independence and shortly afterwards emigrated to Canada where he taught philosophy at Carleton University. In 1955 he moved to the United States and became a professor at the New School for Social Research in New York City. He retired from teaching in 1976 but stayed in New Rochelle, close to New York, until the end of his long life in 1993.

His philosophical reflections on life were published as *The Phenomenon of Life* in 1966. Ever since his emigration, Jonas had been writing in English. In 1979, however, he drafted a book in German again: *The Imperative of Responsibility.*[18] In this work, which turned out to be a major success in Germany, he proposes a new "ethics for the technological age," pointing out the ambiguities and dangers connected with modern technology and criticizing the utopian elements present in the pervasive idea of progress. Given the new situation, in which our acts have global, and at the same time often unforeseeable effects, and given the absolute duty for humankind to exist, we need our technological choices to be guided by a "heuristics of fear."[19] By this latter concept he does not mean timidity or a fear *of* something, but rather a fear *for* something, namely for "the image of man," which we may come to understand better precisely by becoming alert to the dangers to it. "*We know the thing at stake only when we know that it is at stake.*"[20] What is also implied here is the disposition to give greater heed to the predictions of possible harm than to the promises of possible benefit when it comes to evaluating the use of technology.[21] The book has been very influential for the German Green movement,[22] and by now many of its ideas have become so commonplace that one

---

17. This is what Lawrence Vogel suggests in his "Hans Jonas's Exodus," which is his very useful introduction to a posthumous collection of Jonas' essays, edited by Vogel himself: Jonas, *Mortality and Morality*, 1–2: "The life-and-death battle, especially on the Italian front, hardened Jonas's resolve to move beyond the historical inquiries of his student years and develop his own philosophy. Appropriately enough, his musings came to focus on the corporeal, metabolic basis of all life and the struggle of all organisms to maintain their lives in the face of the ever-present threat of not-being or death."

18. Jonas, *Das Prinzip Verantwortung.* English: Jonas, *Imperative of Responsibility.*

19. Cf. for instance, Jonas, *Imperative of Responsibility*, 26–27.

20. Ibid., 27. All italics in citations throughout this book are original.

21. Cf. ibid., 31.

22. Cf. Vogel, "Hans Jonas's Exodus," 3.

may easily forget that it was rather revolutionary when it was published more than thirty years ago.[23] As a sign of his success in Germany, Jonas received the "Peace Prize of the German Book Trade" in 1987, which put him in the illustrious company of thinkers such as Martin Buber, Karl Jaspers, and Gabriel Marcel. Contrary to custom, the award ceremony did not take place in Frankfurt but in Mönchengladbach, Jonas' native city, where, on the same occasion, he also received honorary citizenship and the Federal Republic's Medal of Honor.[24]

In the last stage of his life, Jonas found the occasion to formulate his own "cosmogonic speculations in which decades of thought about ontology and the philosophy of nature found expression."[25] Here he made explicit some of his ideas that had already been more or less implicit in his earlier thought and that regard the questions of the genesis not only of life but of the whole cosmos and the relation of God and the world. In particular, how do we need to think of God, given that he allowed radical evil in his world, such as the horror of Auschwitz and everything which that name stands for?[26] Thus, one can find roughly four phases in Jonas' writing that can be summarized in the following points: (1) Gnosticism, (2) philosophical biology, (3) responsibility and technology, and (4) cosmogonic speculations and theodicy.[27] We will mainly be concerned here with the second and third, even though elements of the first and fourth may also enter occasionally.

In our endeavor to ask about the meaning of human freedom in our biotechnological age in the thought of Hans Jonas, we will broadly proceed as follows. We will dedicate the first part of this book to Jonas'

23. For Jonas' influence on political programs, see, for instance, Schmidt, "Die Aktualität der Ethik von Hans Jonas," 558: "Jonas also influenced political programs. His ethics of responsibility was taken up and made concrete by the Brundtland Commission for the environment and development in 1987. The commission propagated the concept of *sustainable development*" (translation my own).

24. Cf. Jonas, *Memoirs*, 259.

25. Jonas, "Matter, Mind, and Creation," 166.

26. See the collection of his essays *Mortality and Morality*, in particular the articles "The Concept of God after Auschwitz: A Jewish Voice," 131–43 and "Matter, Mind, and Creation: Cosmological Evidence and Cosmogonic Speculation," 165–97.

27. Cf. the categorization that Jonas' wife Lore gives of his work in her Introductory Remarks to his autobiography: Jonas, *Memoirs*, xvi. She finds three elemental phases, corresponding to his three major publications: his book on Gnosis, his *Phenomenon of Life*, and his *Imperative of Responsibility*. In a similar attempt at categorizing Jonas' work, Lawrence Vogel names as the final stage a theological one, which we find justified and would add as a fourth point to Lore Jonas' list (cf. Vogel, "Foreword," xiv).

philosophy of the organism, where freedom is primarily revealed as the freedom of the living being's form with respect to its matter. Then we will turn to Jonas' reflections on our technological civilization, where freedom is revealed to exist in closest conjunction with responsibility. Third, we will argue for the continued relevance of Jonas' thought by examining its influence on a relatively recent and important publication by Jürgen Habermas, who in his *The Future of Human Nature* presents a noteworthy case against genetic enhancement.

1

Hans Jonas' Philosophy of the Organism

IT IS A SIGNIFICANT fact that Jonas' work collecting his "philosophical biology," *The Phenomenon of Life,* is called in its first German edition *Organismus und Freiheit,* "Organism and Freedom,"[1] and indeed the notion of freedom plays a central role in the book, the general tenor of which Jonas already formulated in the letters he sent to his wife from the front while he was serving as a soldier during the War.[2] In one of these letters he writes, "With the concept of freedom, we have a guiding concept for the interpretation of life. . . . In this descriptive sense, freedom is therefore an ontological, foundational character of life as such."[3] In a similar vein, we read in the introductory chapter of *The Phenomenon of Life,* "The concept of freedom can indeed guide us like Ariadne's thread through the interpretation of Life."[4] In what follows we will examine this relationship between life and freedom in Jonas' thought more closely.

1. Jonas, *Organismus und Freiheit.*

2. Cf. Vogel, "Hans Jonas's Exodus," 2: "[Elinore] sent Hans the latest publications in biology, and Hans replied with two sorts of letters: . . . love-letters and teaching-letters. In the latter he sketched ideas that would form the heart of *The Phenomenon of Life,* to be published over twenty years later."

3. Jonas, *Memoirs,* 226 (letter to his wife, dated March 31, 1944).

4. Jonas, *Phenomenon of Life,* 3.

Toward an Ontology of Life

When hearing about a philosophy of the organism, one may think of something very particular and wonder about its relevance for philosophy as a whole. Here it is important to note that Jonas understands his philosophy of the organism not just as a little side-department of philosophy, but as a fundamental approach to ontology: "The problem of life, and with it that of the body, ought to stand in the center of ontology . . . . Life means material life, i.e., living body, i.e., organic being."[5] In this way, following his teacher Heidegger, the question he asks himself is nothing less than "What is being?"[6] Heidegger's approach in *Being and Time* was to turn to the being who is able to ask that question. For Heidegger the analysis of human existence or *Dasein* was the way to get at being-as-such.[7] Yet Jonas points out that his teacher had "forgotten" to consider a very fundamental fact about human persons: they are *living* beings whose mode of existence is *corporeal*.[8]

---

5. Ibid., 25.

6. Cf. Russo, *La biologia filosofica*, 27: "Substantially, the fragments which Jonas, in the form of essays and conferences, dedicates to the interpretation of the living constitute much more than episodes of a regional ontology. They are rather in-depth renderings of a true and proper *system* of the ontology of life as 'fundamental ontology', in a sense that is *analogous* to the meaning that Heidegger has given to this expression in *Being and Time*" (translation my own).

7. Heidegger, *Being and Time*, 32: "Dasein itself has a special distinctiveness compared with other entities, and it is worth our while to bring this to view in a provisional way. . . . Dasein is an entity which does not just occur among other entities. Rather it is ontically distinguished by the fact that, in its very Being, that Being is an *issue* for it. But in that case, this is a constitutive state of Dasein's Being, and this implies that Dasein, in its Being, has a relation towards that Being—a relationship which itself is one of Being. And this means further that there is some way in which Dasein understands itself in its Being, and that to some degree it does so explicitly. It is peculiar to this entity that with and through its Being, this Being is disclosed to it. *Understanding of Being is itself a definite characteristic of Dasein's Being.*"

8. Cf. Hans Jonas' critique of Heidegger in his essay "Philosophy at the End of the Century," 820–21: "Heidegger's concept of *Dasein* as "care" and as mortal is certainly more in keeping with our being's subjugation to nature than is Husserl's 'pure consciousness.' The adjective 'mortal' in particular calls attention to the *existence* of the body with all its crass and demanding materiality. And the world can be 'at hand' only for a being who possesses hands. But is the body ever mentioned? Is 'care' ever traced back to it, to concern about nourishment, for instance—indeed to *physical* needs at all? Except for its interior aspects, does Heidegger ever mention that side of our nature by means of which, quite externally, we ourselves belong to the world experienced by the senses, that world of which we, in blunt objective terms, are a part? Not that I know of."

Putting life at the center of ontology, Jonas remedies this oversight, but still chooses an approach that is similar to Heidegger's inasmuch as he makes use of a "descending ontology." Commenting on Jonas' remark that "man . . . [is] the supreme outcome of nature's purposive labor,"[9] Jan Schmidt notes that "evidently, here Heidegger's descending ontology shines through. Heidegger did not understand nature in an ascending way, beginning from what is elementary to what is complex, as it is customary in the classical-modern natural sciences, but rather in a descending way, from what is complex to what is elementary. Hence, what nature is, is revealed most clearly in the human being—and not reductively in the atom."[10] Asking the same question, then, as Heidegger—where does being reveal itself?—and using a similar method—going from what is more complex to what is more elementary—he nonetheless uses a very different approach by placing the living organism at the center of his reflections. The privileged *locus* of being's appearance is in the living organism, from the amoeba all the way to the human being. The fundamental question of any ontology worth its name thus has to be "What is life?" This is certainly not an easy question, and Jonas makes no pretensions to have answered it in an exhaustive way. Yet his phenomenological reflections may take us a long way and will help us to see what kind of freedom is proper to the organic being and how this freedom is inscribed into its very structure.

Panvitalism

In the opening article of *The Phenomenon of Life*, entitled "Life, Death, and the Body in the Theory of Being," Jonas begins by arguing that the first, in some way "natural" view of the world, held by our ancestors ages ago and perhaps still today enjoying currency among some isolated tribes, was animism or panvitalism: the world is alive.[11] Life and being are coextensive. The sun and the stars are no less alive than these stones, this dust, these plants, along with the lions and tigers and bears. In a world that is alive, death looms in as the great mystery.[12] The corpse is the inexplicable par excellence. The only solution to this mystery is to explain

---

9. Jonas, *Imperative of Responsibility*, 82.

10. Schmidt, "Die Aktualität der Ethik," 565n74, (translation my own).

11. Cf. Jonas, *Phenomenon of Life*, 7.

12. Cf. ibid., 8.

it away: death is not real; it is simply a rite of passage. As Jonas puts it, "Such a negation is the belief in a survival after death which primeval burial customs express. The cult of the dead and the belief in immortality of whatever shape . . . are the running argument of the life-creed with death."[13]

Dualism

At some point, particularly with the great cosmological discoveries, we arrive at a complete paradigm shift. The earth, sun and stars are certainly not alive. In the vast cosmos, in the vast expanses of space that the scientists have discovered, life is the absolute and extraordinary exception. In the context of this thought, as Jonas puts it, "Death is the natural thing, life the problem. From the physical sciences there spread over the conception of all existence an ontology whose model entity is pure matter, stripped of all features of life."[14] The scientific method, which, with its use of analysis and measurement, sets the new standard for knowability, is much more adept at dealing with dead matter than with living things.[15] A living thing does not readily lend itself to analysis and mathematical description. While the lifeless can be readily known, the living is a puzzling and mysterious exception. In fact, "only as a corpse is the body plainly intelligible."[16] In what Jonas calls the "ontology of death,"[17] which is proper to the way modernity looks at the world, the almost all-encompassing rule is dead matter.

In this way, after his discussion of animism, Jonas traces the further development from dualism to two alternative monisms: materialism and idealism respectively. As panvitalism came to be regarded as untenable, an alternative had to be sought. In the search for a consistent world view in which death is the rule, one would have to seek to account for what is alive in terms of what is dead, or at least, in case there is a residual exception, account for latter. This is where dualism comes in. For the validity of any rule, it is convenient to reduce exceptions to the minimum possible. Descartes' efforts to understand animals as mere machines have to be

13. Ibid., 9.
14. Ibid.
15. Cf. ibid., 9–10.
16. Ibid., 12.
17. Cf. ibid., 11.

seen in this context. For him, animals act *as if* they had a certain inward-ness, a certain subjectivity, but in the end they are mere automatons.[18] As Jonas puts it, according to Descartes, "All signs of pleasure and pain in animals are deceptive appearance, i.e., taken for such signs only by an unjustified inference from the habitual connection that in our case obtains between them and certain feelings."[19] For Jonas, "the gain of this *tour de force* lay in its confining the locus of inwardness in nature to the solitary case of man. Puzzling as it was there, it was an exception to the otherwise universal rule and left the rest of living nature free for purely mechanical analysis."[20] In other words, Descartes' idea of the animal au-tomaton helped him to have to admit of only one exception to the univer-sal rule of lifeless matter, i.e., the being whose inwardness is directly given to me, and that is I myself in my own lived experience. Descartes then tried to account for this exception by introducing the split between *res cogitans* and *res extensa,* between the "thinking thing" and the "extended thing," that is, between the mind and the body.[21] In the end, the body is a machine inhabited by a thinking thing, which is the only thing having "inwardness," even though one can no longer properly call it "alive."[22] Life is reduced to consciousness and thus explained away, while conscious-ness is reduced to some mysterious phenomenon that it seems can be bracketed when we are dealing with the scientific examination of the world. Jonas argues that this split of reality into two fields appears to be promising for the scientific treatment of these domains: "We then would have a phenomenology of consciousness and a physics of extension, and the method of one discipline would be as necessarily idealistic as that of the other materialistic. . . . Here the mutual relation of the two seems to be that, not of alternative, but of complementation: 'sciences of nature— sciences of mind.'"[23] It would seem that natural science has neither the

18. Cf. ibid., 41.

19. Ibid., 55.

20. Ibid., 55–56.

21. Cf. ibid., 54.

22. Ibid., 22: "If matter was left dead on the one side, then surely consciousness, brought into relief against it on the other side and becoming heir to all animistic vital-ity should be the repository, even the distillate of life? But life does not bear distillation; it is somewhere between the purified aspects—in their concretion. The abstractions themselves do not live. In truth, we repeat, the pure consciousness is as little alive as the pure matter standing over against."

23. Ibid., 17.

need to affirm nor to deny consciousness. Its object is something else: matter that can be measured, analyzed, dissected and put together again, i.e., dead matter. Whether or not there is consciousness in the world does not appear to be a question natural science needs to ask; it purports to be neutral with regards to the issue.[24]

Jonas maintains that nonetheless natural science cannot uphold this "agnostic" stance about consciousness for very long. Among its objects there is one that forces it to make a choice: the living body, which is both a material being and an entity that exhibits signs of inwardness. The living body testifies to the fact that the two fields ultimately are not separate and that they do not exist in separation from each other. As Jonas puts it, "The fact of life, as the psychophysical unity which the organism exhibits, renders the separation illusory. The actual coincidence of inwardness and outwardness in the *body* compels the two ways of knowledge to define their relation otherwise than by separate subjects."[25]

The main problem of Cartesian mind-body dualism for Jonas is represented by the question of interaction. How do mind and body, which on this hypothesis are two different substances—and two substances of a very different kind—interact with each other? "Cartesian dualism created the riddle of how an act of will can move a limb, since the limb as part of the extended world can only be moved by another body's imparting its antecedent motion to it. Yet after learning from theory that it cannot be, we still go on feeling that we do move our arms 'at will.'"[26] As Jonas says, "Its forte from the point of view of corporeal science, the mutual causal unrelatedness of the two orders of being, was also its mortal weakness (of which 'occasionalism' was the clear confession)."[27] Occasionalism is a philosophical construct that Descartes' followers developed to solve this problem,[28] a problem of which Descartes himself was aware, but which he treated only unsatisfactorily by positing the point of interaction in the

24. Cf. ibid., 54–55.

25. Ibid., 17–18.

26. Ibid., 61–62.

27. Ibid., 55.

28. Cf. for instance the work of Nicolas Malebranche. "Ocassionalism states that all so-called 'second' or 'natural' causes are not true causes at all, but serve merely as occasions on which the true cause (God) operates. . . . Earlier Cartesians such as Cordemoy and La Forge had articulated semi-occasionalist positions, usually denying causal powers to bodies. It is only in Malebranche, however, that we find a full-blooded occasionalism, denying all causal powers also to finite spirits. Only God, for Malebranche, has the power to bring anything about" (Pyle, *Malebranche*, 96).

pineal gland.[29] The pineal gland is of course itself a material reality, so that reference to it is of no avail in solving the issue of how the immaterial mind and the material body can interact. Occasionalism tries to improve on Descartes by arguing that on the "occasion" of every movement of the mind, no one less than God himself supplies the movement of the body. This position is so fantastic that it ultimately amounts to admitting that dualism is untenable. It was of course not a totally new theory at Descartes' time. Thomas Aquinas, writing some four hundred year before him, already had the opportunity of providing a thorough theological and philosophical critique of this idea, which some Islamic philosophers had maintained for other reasons.[30]

Dualism reduces life to consciousness and introduces into our understanding of reality the tension between mind and matter. With these two steps it creates problems that ultimately cause it to founder and to dissolve itself into a monism on either side of its polarity: idealism or materialism. Jonas puts it this way: "In the postdualistic situation there are, on principle, not one but two possibilities of monism, represented by modern materialism and modern idealism respectively: they both presuppose the ontological polarization which dualism had generated, and either takes its stand in one of the two poles, to comprehend from this vantage point the whole of reality."[31] Thus, the options are either to understand everything as consciousness, and to say that what is experienced as matter is just one of its modes, or to claim that everything is matter and that what is experienced as consciousness, nay, the "experience" of experience itself, is just a mode of matter, an epiphenomenon.

Idealism and Materialism

Of the two monisms that are left behind after dualism's demise, idealism is the one to which Jonas gives short shrift. It is indeed a philosophical construct that is internally consistent and as such cannot be meaningfully falsified. But here is precisely its weakness. Though in itself consistent, it is inherently solipsistic. If all reality is ultimately consciousness or

29. Cf. Descartes, *Passions of the Soul*, 41: "And the activity of the soul consists entirely in the fact that simply by willing something it brings it about that the little gland to which it is closely joined moves in the manner required to produce the effect corresponding to this volition."

30. Cf. Thomas Aquinas, *Summa Contra Gentiles*, Book III, Chapter 69.

31. Jonas, *Phenomenon of Life*, 16.

mind, then the question arises of how different minds can be individu-
ated or how they can interact. "Without the self-transcendence of the
ego in *action*, i.e., in the physical dealings with the environment and in
the attendant vulnerability of its being, the closure of the mental order
is logically unassailable, and solipsism can appear as rational discretion
instead of madness."[32] If we begin with consciousness, we will not get
out of consciousness. If all there is, is mind, then there will not be room
for several things; I will always remain trapped within myself. As Jonas
shows convincingly, however, the moment a solipsist argues this position, he or she falls into a performative contradiction, because by its very
nature an argument presupposes the other to whom it is directed and
whom it is meant to convince.[33]

Materialism for Jonas is the intellectually more honest position because, in contrast to idealism, it is open to dialogue and philosophical
discussion inasmuch as it is in principle falsifiable: "Materialism is the
more interesting and more serious variant of modern ontology than idealism. . . . It exposes itself to the real ontological test and with it to the risk
of failure: it gives itself the opportunity of knocking against its limit—and
there against the ontological problem."[34] It is to materialism, then, that
Jonas devotes most of his attention. As a monism, materialism seeks to
account for what we experience as inwardness, mind, or consciousness
in terms of material causes. What seem to be mental states in their own
right, such as anger, joy, love, conviction, and even thought itself, are
nothing but epiphenomena of material states, i.e., phenomena that accompany material states but that themselves have no causal relevance.[35]

Jonas' critique of materialism is twofold. First, he makes a very
original argument that is meant to show how materialism violates some
fundamental rules of natural science. Second, he demonstrates that one
cannot argue for it without contradicting oneself in the act. As to the first,
the phenomenon of the mind is an undeniable reality. We do have the experience of the mind, of thinking, willing, and feeling. Epiphenomenalism accounts for these experiences by claiming them to be byproducts of

32. Ibid., 32.

33. Cf. ibid., 32–33n5: "Not that anyone but a madman has ever taken solipsism
seriously: arguing for it, except in soliloquy, is to acknowledge the 'other' whose consensus is sought. The argument is then frivolous, *qua* dialogue, while the absolute
monologue is the madman's privilege."

34. Ibid., 20.

35. Cf. ibid., 127–28.

material processes. But even as byproducts of material states, these experiences are not nothing—which is shown by the very fact that one tries to account for them. They are *something,* namely, on that theory, the *effects* of material goings-on. However, by being completely reducible to their causes, these effects do not draw any energy from them, nor do they have any energy on their own, which means that they are denied any causal efficacy whatsoever. Thus, here we have the curious case of an effect that is caused without the deployment of energy and that unlike any other known effect is not itself the cause of anything else. This, Jonas argues, is a straight-out contradiction of a fundamental law of physics, namely the law of the conservation of energy. Epiphenomenalism is meant to "denote an effect which, unlike all other effects in nature, does not consume the energy of its cause; it is not a transformation and continuation of such energy, and therefore, again unlike all other effects, it cannot become a cause itself. It is powerless in the absolute sense, a dead-end alley off the highway of causality, past which the traffic of cause and effect rolls as if it were not there at all."[36]

With his second argument against materialism, Jonas does not definitively prove that materialism is false, but he nonetheless convincingly shows that materialism cannot be rationally proposed as a philosophical or scientific theory. By reducing mental states, including thoughts and convictions, to mere epiphenomena of material states such as chemical reactions in the brain, materialism denies the condition of the possibility of rational discourse and even of forming theories in the first place. Even scientists or philosophers who propose materialism as a valid worldview would like to do so because they are convinced that it is true. But if materialism were true, then any given conviction of theirs would be in no relation to its possible truth or falsity. They would simply have to think it true due to some neuronal processes in their brains. And yet, as they advance their position in rational discourse, they are convinced that they hold it true because they have arrived at it after rational reflection and not because they are determined to hold it true by some material reactions in their brains. In other words, materialism denies the possibility of

36. Ibid., 128. Cf. also Jonas' very succinct reformulation of this argument in a later essay that was introduced as appendix in his *The Imperative of Responsibility:* "Epiphenomenalism makes matter the cause of mind and mind the cause of nothing. But causal zero-value is compatible with nothing adhering to matter; and in particular it runs plainly counter to the idea of causal dependency itself that something dependent should be an end only (effect only) and not also in its turn a beginning (a cause) in the chain of determination" (Jonas, *Imperative of Responsibility,* 211).

rational argument. As Jonas puts it, "There is a logical absurdity involved in epiphenomenalism in that it denies itself the status of an argument by depriving any argument of that status. The present argument, no less than that against which it argues, is by this view the epiphenomenon of physical occurrences determined by necessities of sequence entirely foreign to 'meaning' and 'truth.'"[37] By entering in discussion with others and by trying to convince them, one assumes that others are open to rational argument and can very well be convinced (and not simply be caused) to change their positions. If the materialist position were true, however, this could not be the case. One might cause people to change their minds, for instance, by inducing certain substances or by threatening violence, but they could not be "convinced" by argument, unless of course one sees an argument as just another form of physical violence. Then there would be no qualitative difference between an argument and torture for example. On the materialists' own position, their conviction is instilled in them by physical causes; they could not hold their position because it is true but because they are caused to hold it by causes indifferent to questions of truth or falsity. "The only possible reference which the epiphenomenon may have to truth is the accidental agreement of its symbols with facts other than the cerebral facts carrying it, but there is no way on the part of those engaged in the argument, marionettes as they are to those necessities, to evaluate the issue on its merits, and thereby to decide between two alternatives, equal as they are in the factuality of their physical occurrence."[38] For materialists, thinking or arguing cannot have any basis in inwardness and must hence be determined entirely by physiological facts. Their arguments therefore cannot claim any ground of validity. They are like "the Cretan declaring all Cretans to be liars."[39]

Jonas' Attempt at a Solution

With what are we then left to understand the relation between body and mind? Jonas has effectively ruled out both dualism and monism under its forms of idealism and materialism: "In a universe formed after the image of the corpse, the single, actual corpse has lost its mystery. All the more does the one unresolved remainder clash with the universal norm: the

37. Jonas, *Phenomenon of Life*, 129.
38. Ibid., 129–30.
39. Ibid., 134.

living organism, which seems to resist the dualistic alternative as much as the alternative dualism-monism itself."[40] Personally, we would like to note that philosophy at times runs into unanswerable questions because the questions it asks are put the wrong way or are already based on faulty assumptions. The possibility of answering the question raised may well turn around whether we formulate the difficulty in terms of the relation between body and *mind* or whether we set it up in terms of the relation between body and *soul*. It seems that only in the former case we run into true aporia. Putting the question in terms of the relationship between body and soul is difficult, especially if, like Aquinas, and perhaps even Aristotle, we want to hold for the possibility of the (human) soul's survival after the separation from its body in death.[41] This soul, nonetheless, which in the case of humans is an intellectual soul—a soul that also thinks and understands—is the formal principle of the body. Here a solution is certainly difficult, but at least thinkable. The mind considered for itself by Descartes, on the contrary, is a principle that is completely foreign to the body, and to explain the interaction between mind and body will have to be impossible.[42]

What, then, is the direction into which Jonas points us? For him, we need to take seriously the evidence of living things. The organism represented the problem that caused the failure both of monism, in its panvitalist and materialist versions, and of dualism:

> The organic body signifies the latent crisis of every known on-
> tology and the criterion of any future one which will be able to
> come forward as a science. As it was first the body on which,
> in the fact of *death*, that antithesis of life and nonlife became
> manifest whose relentless pressure on thought destroyed the

40. Ibid., 15.

41. Thomas Aquinas, *Summa Theologica* I, 75, 2; cf. also Aristotle, *On the Soul*, I, 1 and III, 5.

42. We can of course grant the pointed observation by Marleen Rozemond that Descartes did not invent dualism: "This problem is often treated as if it was new with Descartes's dualism because his view that the mind is incorporeal is usually approached as if new. But the incorporeity of the mind or the soul was surely not a novelty introduced by Descartes. In the history of Western philosophy it is at least as old as Plato—a fact often ignored in discussions of Descartes's dualism" (Rozemond, "Descartes on Mind-Body Interaction," 435. But whether the problem is formulated in terms of body-mind interaction or body-soul interaction would have to make a difference. Indeed, she goes on to say, "For the Aristotelian scholastics the soul was the form of the body, and in this regard they differed sharply from Descartes" (Rozemond, "Descartes on Mind-Body Interaction," 437).

primitive panvitalism and caused the image of being to split, so it is conversely the concrete unity manifest in its *life* on which in turn the dualism of the two substances founders, and again this bi-unity which also brings to grief both alternatives branching off from dualism, whenever they—as they cannot help doing—enlarge themselves into total ontologies.[43]

Therefore, the organism needs to serve as the measure of any alternative proposal: "The living body that can die . . . is the memento of the still unsolved question of ontology, 'What is being?' and must be the canon of coming attempts to solve it." For Jonas such a proposal needs to go "beyond the partial abstractions ('body and soul,' 'extension and thought,' and the like) toward the hidden ground of their unity and thus strive for an integral monism on a plane above the solidified alternatives."[44] Jonas' search for an "integral monism" that does justice to the phenomenon of life is a quest he set out on in a more systematic manner only toward the end of his life, particularly in the essays published in *Mortality and Morality*. In his article "Matter, Mind, and Creation: Cosmological Evidence and Cosmogonic Speculation," he makes explicit some of his thoughts on the coming to be of the cosmos and the nature of matter and mind that had hitherto been implicit in his thought.[45] In order to think of a monist solution to the problem, which nonetheless takes life seriously, he proposes to replenish and revise the concept of matter "beyond the external qualities abstracted from it and measured by physics; and this means, therefore, a meta-physics of the material substance of the world."[46] This kind of matter must have carried within it from the very beginning, from the "Big Bang," if we want, "an original endowment with the *possibility* of eventual inwardness."[47]

We may wonder then whether at his heart Jonas is not a panvitalist after all. Robert Spaemann and Reinhard Löw claim they can find in Jonas' thought a subtle form of the theory of an organic "world soul," to which even the motion of the anorganic can be traced back.[48] And

43. Jonas, *Phenomenon of Life*, 19.

44. Ibid.

45. Cf. Jonas, *Mortality and Morality*, 166: "I suddenly found myself drawn into my own cosmogonic speculations in which decades of thought about ontology and the philosophy of nature found expression."

46. Ibid., 172.

47. Ibid.

48. Cf. Spaemann and Löw, *Natürliche Ziele*, 31, where, summarizing the panvitalist

indeed, when on different occasions Jonas tells the myth of a divinity that completely surrenders itself in the act of creation and ultimately puts its own destiny into the hands of its creatures,[49] we can get the impression that we are dealing with a type of world soul here.

At the same time, Jonas is critical of Alfred Whitehead's approach in claiming that it does not do justice to the *difference* between animate and inanimate things. Thus, Jonas writes, "Whitehead, who significantly called his general theory of being a 'philosophy of organism,' in effect turned the difference between life and nonlife from one of essence into one of degree."[50] The gravest problem our author sees with Whitehead's take on the issue is that by blurring the distinction between living and non-living beings, Whitehead can no longer make sense of the phenomenon of death,[51] which is to Jonas' mind, however, constitutive of the phenomenon of life. What life is, is fully revealed in its confrontation with death, and an account of life that essentially negates death cannot do justice to life either: "What understanding of life can there be without an understanding of death? The deep anxiety of biological existence has

---

position, they write: "Even the motions of the anorganic, which are never entirely without direction, are traced back to the organic principle of the world soul, which permeates the whole cosmos." In a note to this statement, they say, "This theory has a long tradition that dates back at least to Schelling (*Of the World Soul*, 1798), and which is present, in a most subtle way, also in Jonas" (262n30; translation my own).

In what follows we will make recurrent reference to the work of Robert Spaemann as a philosopher whose thought in many ways echoes Jonas' best insights. The affinity between the two is evidenced by the fact that when Jonas received the Peace Award of the German Book Trade, Spaemann held the *laudatio* (cf. Spaemann, *Schritte über uns hinaus*, 201–13).

49. Cf. Jonas, *Phenomenon of Life*, 275: "In the beginning, for unknowable reasons, the ground of being, or the Divine, chose to give itself over to the chance and risk and endless variety of becoming. And wholly so: entering into the adventure of space and time, the deity held back nothing of itself." The result is that now "God's own destiny, his doing and undoing, is at stake in this universe to whose unknowing dealings he committed his substance, and man has become the eminent repository of this supreme and ever betrayable trust" (ibid., 274). See also: Jonas, *Mortality and Morality*, 131–43 and 188–92. In all these places Jonas insists that his intention is not to develop a systematic doctrine but to tell a myth the point of which is entirely speculative.

50. Jonas, *Phenomenon of Life*, 95.

51. Cf. ibid., 96: "While the polarity of self and world, as also that of freedom and necessity, is taken care of in Whitehead's system, that of being and not-being is definitely not—and therefore not the phenomenon of death (nor, incidentally, that of evil)."

no place in the magnificent scheme."[52] Asked in an interview about the relation between mind and body, Jonas in fact refers to Whitehead, saying that, insofar as he posits feeling and inwardness as primary entities, his position "doesn't make much sense."[53]

And yet again, it is hard to see how his own proposal is essentially different from Whitehead's when he maintains that matter, at its core, has something that is "more" than what natural science can get at. As long as life is not there, this "more" is not necessary for its description; however, matter "must have this something more so that, given the opportunity, life will come forth from matter, and with life will open up a dimension of subjectivity."[54] Now when Jonas speaks about the "given opportunity," he implies some kind of potentiality, which in turn implies the possibility of motion. And here, together with Aristotle, we may ask whether the source of the movement is within the thing that has this potentiality or whether it is outside that thing.[55] In other words, are we speaking about a potentiality like that of an acorn, which, given suitable conditions, becomes an oak tree, i.e., an active potency, or is it the potentiality with which the oak tree, given a suitable outside intervention, can become a chair, i.e., a passive potency? If it is the former kind of potentiality, for which change and becoming is from an inner principle, i.e., from the nature of the thing, the change being the full realization of the thing's nature, then matter already needs to contain some of the reality proper to life, and Jonas' view would not be different from Whitehead's after all. However, if we speak of the latter kind of potentiality, then we will need to posit some extrinsic principle that brings the change about, leading us to fall back into a dualist position.

We see that coming up with a post-dualistic monism that is neither panvitalistic nor materialistic is very difficult, and we will have no ambitions to present or even develop such a position here. Jonas himself,

52. Ibid., 96.

53. Jonas and Scodel, "An Interview," 355: "It doesn't make much sense, though Whitehead did ask exactly that: What do molecules or electrons feel? How do they experience their being? According to Whitehead, they are experiences. Not only do they have them, but they are occasions of feeling. That's Whitehead's formula for the ultimate entities, I think he calls them. The most elementary entities are instances of feeling, and he in that respect comes close to Leibniz's Monadology: that the corporeality is a compound appearance of what is, in its true essence, somehow a mental event."

54. Ibid., 356.

55. Cf. Aristotle, *Metaphysics*, IV, 12.

it seems, did not convincingly manage to do so. Leon Kass comments that, measured against its grand ambition to present a post-dualistic view of life, "one which would again see each organism as a psychophysical unity," but which would at the same time "do justice to the specific difference and unquestionable superiority of the human animal," Jonas' book *The Phenomenon of Life* "can be hailed as only an incomplete success."[56] And also when we look at Jonas' later writings, we cannot detect any substantial step beyond what he proposes there. Nonetheless, as Kass maintains, Jonas does point in the right direction. In fact, on the way to a full philosophical biology that is faithful to the phenomenon of life, Jonas does "establish some major triumphs," as he "succeeds in showing that and how every living organism is a psychophysical unity archetypically concrete, a grown-togetherness of organized outwardly perceivable matter and inwardly experienceable feeling-and-awareness. . . . He shows the necessity of teleologic notions for a true account of life."[57]

To summarize then, did Jonas manage to present a consistent and convincing new ontology that supersedes the aporic alternative monism-dualism? It seems that he did not, even though it is his undoubted merit to have insisted that any such ontology needs to take the testimony of life seriously.[58] Did he develop a complete philosophy of life as a first step toward such new ontology? It seems that to this question, too, we must answer in the negative, even though he did point out some essential features of life, and this is, in the end, all he was setting out to do in the

---

56. Kass, "Appreciating *The Phenomenon of Life*," 4. In our discussion of Jonas, we will make frequent references to Kass, since Jonas was his close friend and the two are among the founding Fellows of the Hastings Center on Bioethics, the first American "think tank" for bioethics (Cf. Kass, "Practicing Ethics," 5–12 and Jonas, *Memoirs*, 200).

57. Kass, "Appreciating *The Phenomenon of Life*," 4.

58. It seems that Robert Spaemann has followed Jonas' lead here, expressing himself in a very balanced way, but without sharing some of Jonas' hesitations: "That which is called 'being' in reference to materiality can in its turn only be understood from the viewpoint of the living. Only being which has the character of being a self is a possible object of benevolence and only for benevolence does being a self reveal itself. When the Psalms bid the sun and the moon, rivers and seas to praise God . . . these are powerful expressions of such universal benevolence, since these creatures praise God by being what they are. But it implies that their being is not merely an objective being-there, but is a tendency, that is, that it is already 'concerned about something,' namely, its own potentiality for being, and that inanimate beings *make possible* something like involvement with themselves and are not just *passively* involved with living beings" (Spaemann, *Happiness and Benevolence*, 101).

first place, as, not without reason, his *The Phenomenon of Life* carries the subtitle _Toward_ a *Philosophical Biology*. He never claimed to develop or to intend to develop a whole system.[59]

The Fundamental Characteristics of the Organism

What then are the fundamental features of Jonas' philosophy of life, and how is life as he describes it related to freedom? As the most defining characteristic of life, Jonas proposes the metabolism. "Metabolism can well serve as the defining property of life: all living things have it, no nonliving thing has it."[60] Life thus takes the form of the metabolizing organism, which is the quintessence of the Aristotelian substance.[61] A connected characteristic mark of living beings is their inner teleology: because of their metabolic existence, "living things are creatures of need. . . . Need is based both on the necessity for the continuous self-renewal of the organism by the metabolic process, and on the organism's elemental urge thus precariously to continue itself."[62] Insofar as they strive to maintain themselves in being, i.e., insofar as for them to be is to live, their being becomes

59. Jonas was not a system-builder. For this also see the testimony of his colleague at the New School, Richard J. Bernstein: "Jonas himself frequently emphasized the ambitiousness and tentativeness of his project. This is why he used *Versuch* and *Search* in the titles of his books" (Bernstein, "Rethinking Responsibility," 19).

60. Jonas, "Burden and Blessing," 34–35.

61. See Robert Sokolowski, who in his "Matter, Elements and Substance in Aristotle," 265n6, gives a short review of several scholars on the matter: "Several recent commentators have observed that simple bodies are not substances 'in the full sense' for Aristotle: H. Cherniss, *Aristotle's Criticism of Plato and the Academy* (New York, 1962), pp. 255, 321, 328 ('living organisms, then, which alone according to Aristotle are in the strict sense substances . . .'); W. Ross, *Aristotle's Metaphysics* (Oxford, 1958), I, p. cxiv ('Living things . . . alone of all perishable things are in the full sense substance'); J. M. Le Blond, *Logique et methode chez Aristote* (Paris, 1939), p. 360, n. 4, puts it well when he says, 'les éléments sont plus des forces que des choses.' J. Owens, 'Matter and Predication in Aristotle,' in *The Concept of Matter*, ed. E. McMullin (Notre Dame, 1963), p. 109, 'Earth, air, and fire, three of the traditional elements, do not seem to him to have sufficient unity in their composition to be recognized as substances.' E. Tugendhat, *Ti kata tinos* (Freiburg, 1958), pp. 84, 94, 97; E. S. Hating, 'Substantial Form in Aristotle's Metaphysics Z III,' *Review of Metaphysics*, vol. 10 (1956–57), p. 311: 'The strong implication of Z is that only living beings, among perishables, and the imperishable heavenly beings . . . are material *ousiai*'; *see* also pp. 708–709."

For a more recent defense of the view that for Aristotle the "primary substances" are living organisms, see Gill, "Matter Against Substance," 379–97.

62. Jonas, *Phenomenon of Life*, 126.

an act and a task. They need to strive actively to keep themselves in existence and to maintain their life. Hence their being is marked by concern, and as such it is purposive, i.e., striving toward ends: "This basic concern of all life, in which necessity and will are bound together, manifests itself on the level of animality as appetite, fear, and all the rest of the emotions. The pang of hunger, the passion of the chase, the fury of combat, the anguish of flight, the lure of love—these, . . . imbue objects with the character of goals, negative or positive, and make behavior purposive."[63] Therefore, it seems that for Jonas the defining characteristics of life are essentially two interconnected ones: the living beings' metabolism and their teleological structure. In what follows we will discuss both of these points and examine how Jonas relates them to freedom.

Metabolism and Freedom

A philosophical account of something as seemingly pedestrian as the metabolism may strike us as odd. Yet Jonas is fascinated by the metabolism and sees in it the very beginnings of freedom. What is it, and what is so special about it? Today it has almost become customary to speak of machines as if they were organisms and to see an analogy between the adding of fuel to a machine and the replenishing of an organism with nutrients, as when someone, requesting a full tank of gas, however jokingly refers to his or her car in endearing terms and says to the attendant at the station, "Just fill her up." Jonas claims that despite the strong prohibitions against anthropomorphism in the realm of nature, it is actually the scientists themselves who are the first to speak about machines in anthropomorphic terms: "Scientists, for so long the very abjurors of anthropomorphism as the sin of sins, are now the most liberal in endowing machines with manlike features."[64] For him, the irony of this state of affairs is "only dimmed by the fact that the real intent of the liberality is to appropriate the donor, man, all the more securely to the realm of the machine."[65] Thus, what we are actually doing is not thinking of machines as if they were organisms, but rather thinking of organisms, including human organisms, as if they were machines.[66]

---

63. Ibid., 126.

64. Ibid., 122.

65. Ibid.

66. In his essay "Life's Irreducible Structure," in which he argues that it is impossible

There are, however, crucial differences between a machine using up its fuel and a metabolizing organism "burning up," if we want, its nourishment. Jonas, arguing against Descartes, makes it clear that a "combustion theory of metabolism" is completely inadequate because "metabolism is more than a method of power generation, or, food is more than fuel."[67] For Jonas these differences are at least two. First of all, the machine, as it is burning its fuel, remains essentially unaltered; the fuel enters the machine, where it is burnt. It leaves the machine's system again in a chemically altered form, but it does not do anything to change the system itself. The only changes that occur within the machine itself are due to normal wear and tear. This wear and tear is an unwanted side-effect that, at least in theory, could be eliminated without altering the machine's functioning; it is not part of the very purpose of the operation. What essentially changes in the operation is the fuel and not the machine.[68]

Things are different with an organism and its metabolism: It is not only the "fuel" or nourishment that changes, but the organism itself. The nutrients enter into its very makeup. Its cells constantly die and are constantly renewed. Indeed, we are what we eat, as nutrition health professionals tend to teach us, and this is the case with every organism. Its cells are constantly changing. Thus Jonas writes about the role of metabolism, "in addition to, and more basic than, providing kinetic energy for the running of the machine (a case anyway not applying to plants), its role is to build up originally and replace continually the very parts of the machine. Metabolism thus is the constant becoming of the machine itself—and this becoming itself is a performance of the machine: but for such performance there is no analogue in the world of machines."[69] In other words, the system that is metabolizing is also the system that results from that process; it is built up and maintained by it. Hence, the object and the agent of metabolism are the same.[70]

Precisely herein lies the phenomenon that Jonas finds so noteworthy: by its metabolism, the living being is the *process* of its own becoming,

to understand living things in terms of physics and chemistry, Michael Polanyi also testifies to this tendency in the natural sciences: "For centuries past, the workings of life have been likened to the working of machines and physiology has been seeking to interpret the organism as a complex network of mechanisms" (1308).

67. Jonas, *Phenomenon of Life*, 76n13.
68. Cf. ibid.
69. Ibid.
70. Cf. ibid.

18

combining identity with constant change. At any given point in time, the metabolizing organism is identical with its matter, but at the next point in time, its matter has already changed, while the organism is still the same. Sylvester is still the same cat today as he was three years ago, even though by now, due to his metabolizing activity, most of his body cells will have changed. But how can we still talk of identity? What is this kind of identity that cannot be based on matter, as it is precisely the material makeup that is constantly changing? Jonas speaks here of the identity of the living form over time: "In this remarkable mode of being, the material parts of which the organism consists at a given instant are to the penetrating observer only temporary, passing contents whose joint material identity does not coincide with the identity of the whole which they enter and leave, and which sustains its own identity by the very act of foreign matter passing through its spatial system, the living *form*. It is never the same materially and yet persists as its same self, *by* not remaining the same matter."[71]

What gives identity to the living being over time is not its material makeup, which keeps changing, but its form—which Aristotle called its "soul" and which has nothing to do with spirits or ghosts as Descartes' idea of the hypostasized soul inhabiting the body may easily suggest.[72] For Aristotle the soul rather refers to the organism's principle of life,[73] that which makes the difference between a living dog and a dead dog, keeping in mind that strictly speaking a dead dog is no longer a dog. On this understanding of the term, not only humans but also cats and dogs and even plants have souls, which is just another way of saying that they are alive. Now, as Jonas expresses himself, this soul or form of the organism enjoys a certain independence from matter, namely with regards to

71. Ibid., 75–76.

72. Contrasting Aristotle's view of the soul with a view of an unnamed source, but easily identifiable as Descartes' taken to its last consequences, Leon Kass writes, "[For Aristotle] the soul was not an ethereal spirit or a ghost-in-the-machine but an immanent and embodied principle of all vital activity" (Kass, *Life, Liberty*, 294).

For an insightful discussion of the notion of "soul" throughout the history of philosophy, see Spaemann, *Persons*, 148–63. For Spaemann, too, just like for Jonas and Kass, the main culprit for the soul's demise is Descartes: "Responsibility for the soul's precarious philosophical status rests chiefly with Descartes, who hypostatized it as an independent soul-substance, united obscurely with a material-substance to compose a human being by their combination. Kant brought weighty arguments to bear against the soul-substance theory, which he accused of 'paralogism'" (Spaemann, *Persons*, 148).

73. Cf. Aristotle, *On the Soul*, II, 1.

this matter.[74] It could also consist of other material parts and it soon will. That the organism now consists of precisely these molecules is accidental to it. It could also consist of other molecules, though of the *same kind,* but *materially different,* and soon it will. Hence we can speak of a freedom of the organism with regards to its matter. In fact, the "metabolism, the basic level of all organic existence, . . . is itself the first form of freedom."[75]

This is the first difference between an organism with its metabolism and a machine with its combustion process: the organism with its living form enjoys a certain freedom in regard to its constituent matter that is unknown to any machine. Now there is a second important difference. Let us suppose a car runs out of gas. It will no longer work. So we put it in a garage and leave it there for a few years until we have saved up enough money to afford another fill. Most likely the car is going to start and work again, and if it does not do so immediately, it will be enough to deal with a few elementary problems, like recharging the battery, and it will be good to go. To put it in more general terms, a machine *can* run on its fuel, but it does not *have to* do so. As Jonas says, "It exists as just the same when there is no feeding at all: it is then the same machine at a standstill."[76] The machine can have extended periods of inactivity, after which it will continue as before. It can *be*—and be inactive. An organism is very different. In its case, the "burning process"—by which it transforms matter into itself—is also a strict and urgent necessity, and hence, in order to maintain itself in being, it has to be active. For living beings, to be is to live,[77] and to live means to be in operation—at the very least: to metabolize. The moment their metabolism stops, they cease to live and hence cease to be. Here Jonas speaks of the "thoroughly 'dialectical' nature of organic freedom," namely the dialectics of freedom and necessity: "Denoting, on the side of freedom, a capacity of organic form, namely to change its matter, metabolism denotes equally the irremissible necessity for it to do so. Its 'can' is a 'must,' since its execution is identical with its being. It can, but it cannot cease to do what it can without ceasing to be."[78] The living beings' freedom is bought at a price, the price of necessity. Their independence from *this* particular matter at any given point in

74. Cf. Jonas, *Phenomenon of Life*, 81.

75. Ibid., 3.

76. Ibid, 76n13.

77. Cf. Aristotle, *On the Soul*, II, 4.

78. Jonas, *Phenomenon of Life*, 83.

time, i.e., their freedom with regards to their constituent matter—which derived from their ability to enter into the metabolic exchange process with their environment—requires them to keep the process operative. The freedom of living things hence is a needful and precarious freedom.

In a very profound reflection, Jonas points out how with the dialectic of the organism's power and need—its ability to stand over and above material nature and at the same time its dependency on that nature—for the first time being appears in an emphatic sense. Only for living things, non-being is a real possibility and a real threat, so that with the appearance of death for the first time being is truly confronted with non-being. It is only the "living substance" that "by some original act of segregation, has taken itself out of the general integration of things in the physical context, set itself over against the world, and introduced the tension of 'to be or not to be' into the neutral assuredness of existence."[79] Only here, in the confrontation with possible non-being, being "assumes an emphatic sense: intrinsically qualified by the threat of its negative it must affirm itself, and existence affirmed is existence as a concern."[80] Insofar as organisms are not self-sufficient, their freedom is a precarious freedom. It is from their environment, which may or may not be accommodating, that they have to retrieve the material sustenance which they need for their survival, a survival that thus itself becomes a task and hence a concern.

It is important to note that for Jonas the organism's "existence as concern" is not only marked by the dialectics of freedom and necessity, but also by a transcendence and relationality. Jonas puts it this way, "Life is essentially relationship; and relationship as such implies 'transcendence,' a going-beyond-itself on the part of that which entertains the relation."[81] The organism is to an extent free from this matter and hence in need to be in constant exchange with matter for which it must go out of itself and be in constant contact and interchange with its environment. Because of its freedom, the organism is needful, and because of its need, it is relational and self-transcending.

---

79. Ibid., 4.

80. Ibid. We note how here Jonas is enlarging the Heideggerian concept of "concern" as the mode of existence not only proper to human *Dasein* but proper to the entire realm of the living.

81. Ibid., 4–5.

Natural Teleology

THE REASONS FOR ITS REJECTION

As we have talked about the metabolism and the living being's need to maintain itself, together with its active concern to do so, we have already implied a concept that has fallen in ill-repute in biology and the rest of the natural sciences, and that is natural teleology. By speaking about the living being's mode of existence as concern, we presuppose that it is a being that takes interest in things, that has ends and purposes, which implies that these ends in turn have effects on it. The living thing is affected by its ends. In other words, if it has ends, these influence its behavior and function as *causes* in the form of causality that Aristotle called "final." For him, final causality was among the four causes, and indeed the most noble and principal one.[82] Things are moved by their ends. For the Philosopher, the whole world is ultimately moved by God, but not in the way a clock is moved by the clockmaker but in the way the lover is moved by the beloved.[83] Although God himself is unmoved, he is the mover of the entire universe as its final cause, the ultimate end toward which all things tend. The living things' efforts to maintain themselves and their species in being can thus ultimately be interpreted as the expression of their desire to participate in the divine.[84] Still in medieval times the world was seen as intrinsically purposive. The presence of ends and purposes in created nature served Thomas Aquinas, for instance, as premise in the formulation of his fifth way. All things act for their ends. Even the falling stone is said to tend toward the center of gravity as its natural end, a tendency that Aquinas refers to as "natural love."[85] With all this purpose and meaning

82. Cf. Spaemann and Löw, *Natürliche Ziele*, 60: "The final cause holds the primacy in the Aristotelian doctrine of causes. It alone can respond to the question 'why' in the satisfactory sense of leading to understanding" (translation my own).

83. Aristotle, *Metaphysics*, XII, 7: "The final cause, then, produces motion as being loved, but all other things move by being moved."

84. Aristotle, *On the Soul*, II, 4: "The acts in which it [the nutritive soul] manifests itself are reproduction and the use of food—reproduction, I say, because for any living thing that has reached its normal development and which is unmutilated, and whose mode of generation is not spontaneous, the most natural act is the production of another like itself, an animal producing an animal, a plant a plant, in order that, as far as its nature allows, it may partake in the eternal and divine. That is the goal towards which all things strive, that for the sake of which they do whatsoever their nature renders possible."

85. Cf. Thomas Aquinas, *Summa Theologica*, I-II, q. 26, art. 1: "The connaturalness

in the world, there had to be a being that has put this meaning there, and this is the being we call God.[86]

Things are largely looked at in a different way today. Whether or not we are inclined to admit of final causes will largely depend on our understanding of knowledge, and indeed the idea of knowledge has since changed. For the ancients and medievals, knowledge was a kind of union between the knower and the known in analogy to the intimate union between a man and a woman. Thus, "Adam knew Eve his wife" and she conceived and bore a son (Gen 4:1 *RSV*). If we understand knowledge in this sense, that is, in the sense of *theoria,* the speculative beholding of eternal truths which unites us to them, then there is room for final causes. In fact, very often it will be precisely these final causes that represent the worthiest objects of our knowledge. Ultimately, the highest cause will be Love itself, so that Dante can speak of "the love that moves the sun and the other stars."[87]

To modern ears these considerations will sound awkward. They raise the suspicion of anthropomorphism, which attributes the same inclinations and tendencies to nature that would seem to be proper to the human being alone. Today's scientific common sense is that only humans act for ends.[88] Nature has no ends of itself, and to read such ends

---

of a heavy body for the centre, is by reason of its weight and may be called 'natural love.'"

86. Cf. Thomas Aquinas, *Summa Theologica*, I, q. 2, art. 3: "We see that things which lack intelligence, such as natural bodies, act for an end, and this is evident from their acting always, or nearly always, in the same way, so as to obtain the best result. Hence it is plain that not fortuitously, but designedly, do they achieve their end. Now whatever lacks intelligence cannot move towards an end, unless it be directed by some being endowed with knowledge and intelligence; as the arrow is shot to its mark by the archer. Therefore some intelligent being exists by whom all natural things are directed to their end; and this being we call God."

87. See the very enlightening article by José Granados, "Love and the Organism," 435–69, who begins his essay precisely with this citation from Dante Alighieri, *Paradiso*, XXXIII, 145.

88. Spaemann and Löw in *Natürliche Ziele*, 252–53, evidently following Jonas here, have aptly pointed out the implausibility of this position, even on scientific grounds, since on its terms, "the dimension of finality suddenly falls from the sky, so to speak, with the human being, thus separating human beings from their natural context. The alternative is thus the following: Either human purposive action is itself ontologically secondary, that is, ultimately a product of an accidental constellation of deterministic causal processes marked by selection . . . or the categorical structure of striving-for is—with various stages of complexity—constitutive of natural beings in general" (translation my own).

into nature is to be unscientific. Jonas summarizes what he conceives to be the modern attitude in these words: "The one reminder deemed sufficient to compromise teleology for the intelligent reader is that final causes have relation to the nature of man rather than to the nature of the universe—implying that no reference must be drawn from the former to the latter. . . . Anthropomorphism at all events, and even zoomorphism in general, became scientific high treason."[89] He continues by arguing that teleology was not abandoned because it was proven false. "Regarding final causes, we must observe that their rejection is a methodological principle guiding inquiry rather than a statement of ascertained fact issuing from inquiry."[90] The motivation for placing a scientific *anathema* on the search for final causes was simply based on the fact that apparently it is not productive, or, in the words of Francis Bacon, that it "is barren . . . like a virgin consecrated to God."[91] For Bacon, the knowledge that rain falls in order to water the farmer's fields and make the crops grow is fruitless because it helps us neither to produce nor to predict rainfall.[92] It may in fact be positively dangerous inasmuch as it feeds into human laziness, which is too easily content with such "useless" explanations and thus keeps us from asking any further.[93] For Bacon, true knowledge had to be "fruitful,"

89. Jonas, *Phenomenon of Life*, 35.

90. Ibid., 34.

91. Bacon, *Dignity and Advancement of Learning*, 512. Cf. Spaemann and Löw, *Natürliche Ziele*, 11.

92. Aristotle did not actually think that rain fell to make the crop grow, for the simple reason that the same rain that makes it grow can also spoil it, namely when the rain falls on it on the threshing floor (cf. *Physics*, II, 8). In his commentary on the *Physics*, Thomas Aquinas criticizes Aristotle on this point for comparing "a general cause with a particular effect." As a general cause rain is the necessary condition for the crop to grow, which is also what generally happens. The spoiling of the crop on the threshing floor, in contrast, is incidental. Even if we agree that the rain does not fall in order to destroy the crop, we do not need to deny that the rain falls in order to make it grow. In other words, the water-cycle as such can still have a teleological nature: there can still be a purposive relation between the rain as such and the thriving of vegetation in general, even if on occasion, the rain may actually harm a plant (cf. Thomas Aquinas, *Commentary on Aristotle's Physics*, Book 2, Lecture 12, n. 5: "For a universal cause is referred to a particular effect. And it must also be noted that the growth and conservation of growing things on earth occur in most cases because of the rain, whereas their corruption occurs in few instances. Hence, although rain is not for their destruction, it does not follow that it is not for their preservation and growth." For this discussion, see Spaemann and Löw, *Natürliche Ziele*, 254).

93. Cf. Jonas, *Phenomenon of Life*, 34: "The mere search for them was quite suddenly, with the inauguration of modern science, held to be at variance with the

that is productive. Hence his famous saying that knowledge is power.[94] What he refers to here is ultimately the power to produce.[95]

How did this fundamental break occur? How did we get from a knowledge in the sense of *theoria*, the intimate union of the knower with the things known, to a knowledge in the sense of production, which no longer looks at things as for what they are but as for what one can do with them? In fact, it seems that at some point at the beginning of modernity people decided that the only things they can know are those they have made themselves, so that knowledge by its nature had to become "fruitful" or productive. In a striking hypothesis, which she bases on an observation made by Alfred North Whitehead, Hannah Arendt links this paradigm shift to a seemingly small and innocent event, which, according to her, would nonetheless forever change the way we look at ourselves and the world around us: the invention of the telescope. "'Since a babe was born in a manger, it may be doubted whether so great a thing has happened with so little stir.' These are the words with which Whitehead introduces Galileo and the discovery of the telescope on the stage of the 'modern world.' . . . These first tentative glances into the universe through an instrument, at once adjusted to human senses and destined to uncover what definitely and forever must lie beyond them, set the stage for an entirely new world."[96]

With this little device Galileo Galilei demonstrated that the earth revolves around the sun and not vice versa. As a theoretical model this idea of course had existed before. But Galileo may indeed claim for himself the achievement of publicly demonstrating this theory to be a fact.[97]

---

scientific attitude, deflecting the searcher from the quest for true causes."

94. Cf. Bacon, *New Organon*, Book I, aphorism 3, 33: "Human knowledge and human power come to the same thing." Cf. also Bacon, *Meditationes sacrae*, 79: "Nam et ipsa scientia potestas est."

95. For a criticism of modern epistemology as based on a false ontology, cf. Schindler, *Heart of the World*, 263–65.

96. Arendt, *Human Condition*, 257–58. Arendt cites from Whitehead, *Science and the Modern World*, 12.

97. Cf. Arendt, *Human Condition*, 259–60: "Ideas, . . . as distinguished from events, are never unprecedented, and empirically unconfirmed speculations about the earth's movement around the sun were no more unprecedented than contemporary theories about atoms would be if they had no basis in experiments and no consequences in the factual world. What Galileo did and what nobody had done before was to use the telescope in such a way that the secrets of the universe were delivered to human cognition 'with the certainty of sense-perception'; that is, he put within the grasp of an earth-bound creature and its body-bound senses what had seemed forever beyond his

For Arendt, this discovery was the cause of an epistemological crisis in educated Europe. It proved that the senses had deceived us for ages, so that they can no longer be trusted to bring us into contact with reality.[98] They tell us that the sun revolves around the earth, while with the telescope Galileo was able to prove that the opposite is true. Who knows if our senses can be trusted in other fields? Someone who lied once can no longer be trusted, so that suspicion of our senses became widespread. Knowledge could no longer be perceived as a somewhat mystical, spiritual union between the knower and the object known. Such a notion of knowledge presupposes that the object known is sincere and intelligible. With a deceitful nature, no mystical union is possible. At most, the human person can become the laughingstock of a material world that is out to fool him.

In addition, Galileo's use of the telescope did not only disprove the sun's presumed rotation around the earth. Jonas, who refers to the same event as Arendt, points out how Galileo had demonstrated the stars and planets around us to be just like the earth. The "conception that nature is the same everywhere, be it heaven or earth," was entirely novel and indeed shattering.[99] He continues: "The earth had become a 'star' itself, viz. a planet, and by the same token the planets had become 'earths.' Instead of enjoying a nobler, more refined and sublime type of being, they were instances of the same physical reality as the one we are familiar with on this gross, material, heavy earth. Thus, with one stroke, the essential difference between the terrestrial and celestial spheres, between sublunar and stellar, corruptible and incorruptible nature vanished: and with this, the idea of *any* natural order of rank lost its most telling support in the

reach, at best open to the uncertainties of speculation and imagination."

Contrary to his own claims, Galileo did not invent the telescope himself. Rather, most likely it was the Dutch spectacle maker Zacharias Janssen from Middelburg. Two years prior to Galileo, the invention was already publically exhibited at the Frankfurt Fair of 1608. The fact however remains that he caused a great *public* stir by publishing the findings to which he came by its use. For all this and for an unusual perspective on the role of the Roman Inquisition in the Galileo case see: Zander, *Kurzgefasste Verteidigung der Heiligen Inquisition.*

98. Cf. Arendt, *Human Condition*, 262: "If we wish to put this into historical perspective, it is as if Galileo's discovery proved in demonstrable fact that both the worst fear and the most presumptuous hope of human speculation, the ancient fear that our senses, our very organs for the reception of reality, might betray us, and the Archimedean wish for a point outside the earth from which to unhinge the world, could only come true together."

99. Jonas, *Philosophical Essays*, 52.

visible scheme of things."[100] Here we are dealing with a true revolution, the "spectacular impact" of which, as Jonas emphasizes, is difficult for us to appreciate properly, since we "have long been accustomed to thinking of 'heavenly' bodies in no other way and have actually seen men walking on the moon."[101] But for people at the time of Galileo, a comprehensive worldview, which was built on the qualitative difference between the earthly and the heavenly spheres, simply collapsed. It is thus not hard to see how Galileo's discovery should have brought with it an epistemological crisis.

A symptom of this crisis is modernity's obsession with apodictic certainty, best represented in the figure of Descartes and his method of systematic doubt. So afraid of being fooled by nature or by some malicious genius, he equates knowledge with certainty: only certain and indubitable knowledge deserves that name.[102] Descartes' quest for certainty leads to curious results. While previous ages sought and found the footprints of God in the material world, taking the things created as evidence for the existence of the Creator, Descartes, no longer sure of the existence of the material world—a world, after all, that is given to him by means of his senses, whom he has learned to mistrust—feels the need to prove the existence of the material world with the help of the idea of God,[103] of whose existence, in turn, he his apodictically certain because of the ontological proof.[104] In contrast to Aquinas' five ways, which are cosmological, that is, argue from the world to God, the ontological proof is a logical one. Thus we see where Descartes places certainty and hence knowledge, namely in consciousness. Logical relations, the constructs of consciousness, conveniently yield certainty without any need for excessive ontological commitments to the real existence of anything. Two apples and two apples are four apples, whether they are real or dreamt, a fact which "neither God nor an evil spirit can change."[105] Much of modern

---

100. Ibid., 52–53.

101. Ibid., 53.

102. Cf. Descartes, *Correspondence*, 147: "Knowledge is conviction based on a reason so strong that it can never be shaken by any stronger reason" (letter to Regius, 24 May 1640).

103. Cf. Descartes, *Meditations on First Philosophy*, Sixth Meditation, 55.

104. Cf. ibid., Fifth Meditation, 46–47.

105. Cf. Arendt, *Human Condition*, 284: "The famous *reductio scientiae ad mathematicam* permits replacement of what is sensuously given by a system of mathematical equations where all real relationships are dissolved into logical relations between

philosophy will indeed follow Descartes in this turn to consciousness, and also the elevation of mathematics to the queen of the sciences can be seen to have one of its roots right here.

However, some philosophers and scientists, who were no less affected by the epistemological crisis occasioned by Galileo, were nonetheless much more practical than Descartes. Instead of placing the blame for the difficulties at the door of our senses, they rather charged nature with it and also suggested a solution. If nature does not freely reveal her secrets, she has to be tricked out of hiding by means of instruments.[106] This new kind of theory, Jonas writes, "achieves knowledge of nature's laws of action by itself engaging nature in action—that is, in experiments, and therefore on terms set by man himself."[107] Experiments are always a practical testing of how a given thing will act and react under certain conditions, and hence to know a thing ultimately means to be able to imagine "what we can do with it when we have it," as Thomas Hobbes puts it.[108]

Robert Spaemann and Reinhard Löw point out that Aristotle also knew of the experimental method, but that it did not seem very helpful to him, since knowledge was supposed to bring us into contact with the way things are by themselves without human intervention: "For him physics is something like behavioral research."[109] In contrast, the questions to which our experiments answer are always set in terms that we ourselves choose and impose. Hence also Hannah Arendt's suspicion that at the end of the day modern scientists do not meet nature but only themselves: "We may . . . fall prey to the suspicion that what we have found may have nothing to do with either the macrocosmos or the microcosmos, that we

man-made symbols. It is this replacement which permits modern science to fulfil its 'task of *producing*' the phenomena and objects it wishes to observe. And the assumption is that neither God nor an evil spirit can change the fact that two and two equal four."

106. Cf. Bacon, *New Organon*, Book I, aphorism 98, 81: "The secrets of nature reveal themselves better through harassments applied by the arts than when they go their own way."

107. Jonas, *Phenomenon of Life*, 190.

108. Cf. Hobbes, *Leviathan*, 17: "The train of regulated thoughts is of two kinds; once, when of an effect imagined, we seek the causes, or means that produce it; and this is common to man and beast. The other, when imagining any thing whatsoever, we seek all the possible effects, that can by it be produced; that is to say, we imagine what we can do with it, when we have it."

109. Spaemann and Löw, *Natürliche Ziele*, 50 and 264n21, (translation my own).

deal only with the patterns of our own mind, the mind which designed the instruments and put nature under its conditions in the experiment . . . in which case it is really as though we were in the hands of an evil spirit who mocks us and frustrates our thirst for knowledge, so that wherever we search for that which we are not, we encounter only the patterns of our own minds."[110]

On modern terms, ultimately, the perfect way of knowing something would be knowledge not only of its behavior under certain conditions but knowledge about the conditions of its coming to be. The perfect experiment in this sense would be the production of the thing itself. To use Jonas' example: to know what a cosmic nebula is, is to be able—at least in theory—to build one if one were sufficiently powerful and had the necessary components at one's disposal.[111] Hence derives the implicit conviction that we can only know what we have made ourselves, or in Jonas' words, "Only a *factum*—what has been made—can be a *verum*."[112] Given this new approach proper to modern science, knowledge, for it to be genuine, needs to have some application. Therefore, in modernity science and technology go hand in hand. Thus, the second way of reacting to the modern epistemological crisis is to reduce knowledge to power: knowledge, i.e., science, is power and in particular the power to produce.[113] By its very method, knowledge has become practical.

If we look at today's science, we notice that in some ways it has managed to unite both ways of responding to the modern epistemological crisis, even though there is no inherent necessity that the *reductio scientiae ad mathematicam* and the *reductio scientiae ad potentiam* should reveal themselves to be one and the same thing. It just turned out that

---

110. Arendt, *Human Condition*, 286–87.

111. Cf. Jonas, *Phenomenon of Life*, 203–4.

112. Ibid., 202.

In this context Hannah Arendt points out how the conviction that we can only know what we have made ourselves—consequent to Cartesian doubt—was at the root of Giambattista Vico's interest in history: "One of the most plausible consequences to be drawn from Cartesian doubt was to abandon the attempt to understand nature and generally to know about things not produced by man, and to turn instead exclusively to things that owed their existence to man. This kind of argument, in fact, made Vico turn his attention from natural science to history, which he thought to be the only sphere where man could obtain certain knowledge, precisely because he dealt here only with the products of human activity" (Arendt, *Human Condition*, 298).

113. Cf. Bacon, *New Organon*, Book I, aphorism 3, 33: "Human knowledge and human power come to the same thing."

mathematics is much more practical than one might have expected. Thus, for instance, we found out that numbers in the form of an mp3-file can reproduce sounds much better than magnetic tapes were ever able to do; numbers can create pictures, sounds, and movies, a digital world, a virtual reality. Since the deciphering of the human DNA, it seems that we can even express what the human person is by means of numbers. Science and technology, knowing and doing, are today more and more intertwined. Since science needs the experiment to verify its hypotheses, since the experiment is the way to knowledge, knowledge can no longer be thought of without application. Knowledge for the sake of knowledge, knowledge in the sense of *theoria* as the awe-filled beholding of eternal truths the way the ancients knew it, has become a contradiction in terms. Science has been reduced to mathematics, and even mathematics has found a powerful application, so that in the last analysis, knowledge is the power to do and in its perfection the power to create—our knowledge has become "creative."[114]

If such, then, is the ideal of knowledge, it is clear that there is no room for teleology, the knowledge of ends. It is the scientist who sets the ends. As Jonas says, "The very process of attaining knowledge leads through manipulation of the things to be known,"[115] and not the observation of their own inner dynamics, so that "the nature of things is left with no dignity of its own."[116] A consideration of any presumed ends inherent to the things the scientist seeks to manipulate would only obstruct the scientific process and would be useless for its results. Nature, once considered *mater* or "mother," now is considered mere *matter,* and matter is finally reduced to *material* for experimental production.[117] As we

114. It is certainly noteworthy that for the ancients and medievals creative knowledge was precisely the prerogative of God: he knows things by creating them. For an analysis of Giambattista Vico's epistemology, which is equating the true with the made, and for the analogy that he establishes between divine and human knowledge, cf. Botturi, *Tempo, linguaggio e azione*, 32–41.

115. Jonas, *Phenomenon of Life*, 205.

116. Ibid., 192.

117. Cf. John Paul II, *Evangelium vitae*, n. 22: "Nature itself, from being 'mater' (mother), is now reduced to being 'matter', and is subjected to every kind of manipulation." Also see Hannah Arendt's remark about the dubious wisdom proper to the human being as producer or *homo faber*: "It can result, and has often enough, in the misrepresentation of all naturally given things as mere material for the human artifice—as though trees where nothing but potential wood, material for tables" (Arendt, "Eggs Speak Up," 283). In other words, the tree (nature) is reduced to its wood (matter), which in turn is reduced to lumber (material).

already mentioned above, the demise of natural teleology was not caused by arguments or demonstrations. Here we can see, however, why giving up its consideration could be regarded as convenient: to think of the inner dynamic of the thing itself can only be distractive in a context where one wishes to introduce it into dynamics set by oneself. Nonetheless, Jonas points out that there is a price to pay for disregarding or denying teleology. For him, one problem has to do with the fate of efficient causality—which in principle is treasured highly in the natural sciences—and another difficulty refers to the fate of the scientists themselves and all other human beings, insofar as the ban on teleology will inevitably have to be extended to human persons as well. To see ends inherent in nature is to see nature anthropomorphically. But, as Jonas notices, "so radically has anthropomorphism been banned from the concept of nature that even man must cease to be conceived anthropomorphically if he is just an accident of nature. As the product of the indifferent, his being, too, must be indifferent."[118] In short, as Spaemann puts it, human beings are in danger of becoming their own anthropomorphism.[119] In the following paragraphs we will deal with both of these problems in more detail.

THE END OF EFFICIENT CAUSALITY

In an intriguing analysis, Jonas makes the case that the fight against final causality in scientific discourse has also had its effects on efficient causality. "The struggle against teleology is a stage in the struggle against anthropomorphism which by itself is as old as Western science. . . . But the argument once under way did not stop there: it overtook even the efficient causation in whose favor final causes had been ousted."[120] While it is true that final causality cannot be measured or observed, the same holds true also for efficient causality. Jonas grants to David Hume that he is right in arguing that we never *perceive* a cause. "Hume has shown

118. Jonas, *Phenomenon of Life*, 233.

119. Cf. Spaemann, "Kommentar," 72–73: "When human beings, even in their most spiritual acts, tentatively become the objects of biological, and specifically of evolutionary hypotheses, this implies that their status as subjects disappears. Human beings in their human self-understanding as free, knowing, and acting beings, i.e., as persons, become their own anthropomorphism. Viewed 'scientifically' they are the self-reproducing states of a material substratum—'beyond freedom and dignity' as the American psychologist Skinner proclaimed" (translation my own).

120. Jonas, *Phenomenon of Life*, 36.

that 'causation' is not found among the contents of sense perception. This is incontrovertible so long as 'perception' is understood, with Hume, as mere *receptivity* that registers the incoming data of sensation."[121] What we *perceive* are two consecutive events. We flip a switch in the room (event A) and the light goes on (event B). We witness both event A and event B; however, we do not witness any "cause" anywhere. From where does the idea arise that A is the cause of B? According to Hume, "after a repetition of similar instances, the mind is carried by habit, upon the appearance of one event, to expect its usual attendant, and to believe that it will exist."[122] After perceiving two events following each other with a certain regularity, the mind, habituated to the sequence, posits the prior event as the cause of the latter. "This connexion, therefore, which we *feel* in the mind, this customary transition of the imagination from one object to its usual attendant, is the sentiment or impression from which we form the idea of power or necessary connexion."[123] Habituation alone, then, is at the origin of our idea of causality: "Contemplate the subject on all sides; you will never find any other origin of that idea."[124] As Jonas puts it, what Hume implies is that the idea of causality, which includes "the idea of force and necessity," is "alien to the record of things."[125] It is rather something that we read into things, and which therefore represents "another case of that transference of traits from human self-experience into nature which had become anathema to objective science."[126] In other words, the ideas of force and necessary connection exclusively belong to human experience—or really, human sentiment—and reading them into nature is another case of that anthropomorphism that science ousted by putting a ban on final causes.

What is the problem here for science? If it accepts these terms, it will resign itself to merely describing things instead of explaining them. Thus, the goal of knowledge will no longer be understanding, so that

---

121. Ibid., 26.

122. Hume, *Enquiry*, 69. It is in this context that Hume gives his famous example of the billiard balls: "The first time a man saw the communication of motion by impulse, as by the shock of two billiard balls, he could not pronounce that the one event was *connected*; but only that it was *conjoined* with the other. After he has observed several instances of this nature, he then pronounces them to be *connected*" (ibid.).

123. Ibid.

124. Ibid.

125. Jonas, *Phenomenon of Life*, 36.

126. Ibid.

modernity's quest for knowledge ultimately abolishes itself. Jonas argues this way, "Discarding the explanatory concept of force as anthropomorphic, and as unverifiable by a mere measuring account of extensity, science restricted its claim to that of registering sequences of positions in a space-time system of coordinates and of formulating quantitative regularities in such sequences as 'the laws of nature.' Explanation has thus been forsaken for mere description."[127] Our journey, which went from panpsychism over dualism to materialism, thus "ends in an agnostic renunciation of the idea of knowledge as an understanding of its objects."[128]

Agreeing with Hume that causality is not a percept, Jonas nonetheless strikes at the heart of Hume's thought to show its grave inner inconsistency, which bespeaks the fact that the ghostly notion of causality cannot be dispelled with so easily. It is an observable and undisputable fact, which Hume does not deny, that people do have the idea of causality. As we have pointed out above, Hume wants to argue that this notion, however, is an illusion and, as we have said, feels the need to explain it away by arguing that what *seem* to be cause and effect are simply two events which the mind (mistakenly) connects in causal terms because it is habituated to their sequence. Succinctly summing up Hume's position as the account of causality's "bastard birth," namely as deriving from "the unsupervised liaisons (mutual attractions) between our ideas," Jonas argues that this approach "does not stand scrutiny."[129] Ironically and inconsistently, Hume's account for our having the illusion of causality, of which he wants to free us, is *itself* an account that uses causal terms: the—supposedly illusionary—notion of causality is *caused* in the mind by habituation: "The 'force of habit' acting in the association of ideas may well, for its own explanation, require to call on physical causes (e.g., brain mechanisms), that is, on the very reality of that which it merely seemed to fake."[130] Pointing out this inconsistency as such does not prove that Hume is wrong when he denies causality as a relation among things; it only shows that his theory does not succeed in accounting for our *idea* of causality—which is a fact of our experience—without presupposing *actual* causality in things, which, again, it tries to deny. It will not help to reduce causation to a relation among ideas in order to avoid having to

127. Ibid.
128. Ibid., 37.
129. Ibid., 26–27.
130. Ibid., 27n1.

speak of it as a relation among things, simply because ideas themselves are also "things" in the widest sense. It is simply inconsistent when, as Jonas puts it, "in their own interrelation [our ideas] are allowed the very dynamics they deny to the things in their portrayal of them."[131] Hume's inability to explain in a consistent way an important fact of our experience, namely the fact that we do have the idea of causation, speaks against his argument.

As we have said, Jonas agrees with Hume that causality is nothing that we see or perceive. "Whichever causality it be, on this point Hume's critique was right that it is not met with any perception, and that the nexus between data is not a datum itself—not a perceived content."[132] However, to look for causes as things to be observed by the senses is to look for them in the wrong place. From where do we get our idea of causality, then? For Jonas it derives from the experience of ourselves as living, striving organisms who have to exert and overcome force to reach their ends. According to him, Hume, and incidentally also Kant, in their puzzlement about causation, "forget the body," assuming "that there is no firsthand knowledge of force, transitivity, and the dynamic bonds of things."[133] It is as bodily beings who have ends and pursue goals that we have to exert force to overcome resistance, thus experiencing ourselves as efficient causes. As Jonas formulates it, "the primary aspect of causality is not regular connection, not even necessary connection, but force and influence."[134] Force and influence in turn are original experiences and not percepts or "interpolations between contents of experience," and their source is "not sense perception, but our body exerting itself in action."[135] Contrary to Kant, then, who held that causality was an *a priori* basis of experience, Jonas argues that causality is itself a basic experience, an experience, which "has its seat in the *effort* I must make to overcome the resistance of worldly matter in my acting and to resist the impact of worldly matter upon myself."[136] This effort is one that I have to exert with and in

131. Ibid., 26–27.

132. Ibid., 25–26.

133. Ibid., 27–28.

134. Ibid., 33.

135. Ibid.

136. Ibid., 23. If the experience of force is at the basis of our idea of causality, then clearly Hume is right when he says that causality is not percept. Jonas writes about force that it is "not a 'datum' but an 'actum,' [it] cannot be 'seen,' i.e., objectified, but only experienced from within when exerted or suffered" (ibid., 31).

my body. From this experience, I then extrapolate "the dynamic image of the world—a world of force and resistance, action and inertia, cause and effect."[137] Jonas sums up his fundamental idea of causality in these terms: "Causality is not the a priori of experience in the understanding but the universal extrapolation from propriobodily prime experience into the whole of reality."[138] He does not voice himself on the issue whether this extrapolation is legitimate but recommends the question to the study of a philosophy of the organic.[139]

For Vittorio Hösle, Jonas' reflections on causality are not convincing: "It may be true that we form the concept of causality only due to the experience of a resistance against our body, but this applies only to the genesis of the concept, not to its application."[140] Hösle seems concerned that on Jonas' account causality may be limited to physical causes, while it should also include psychic ones: "To my mind, Kant is correct when he construes a concept of causality that is more universal than the one proposed by Jonas, inasmuch as it includes physical and psychical causes."[141] In our view, Jonas does not seem to prejudge the matter. As Hösle himself grants, what Jonas is primarily interested in is to show where our notion of causality comes from and not to ask what kind of entities it applies to. To raise the question of the relation between physical and psychic causes in this context is to reopen the whole question of an adequate account of body and soul in post-dualist, monist terms that are neither materialistic nor idealistic and for the sake of which Jonas points us toward the phenomenon of life, but which he himself does not wish to develop in more detail at this particular point.

If Jonas is right with his alternative pedigree of efficient causality, then we can see how intimately it is connected with final causality. The experience of being efficient causes is rooted in our bodily existence. As bodies, we experience the force we exert on things together with their resistance, which we seek to overcome. But once we speak about force and resistance, we imply the notion of effort, which in turn involves the idea of striving-for. We only make efforts *for the sake of* some goal. Without any goal, we would just stay put. Hence, we only act and experience ourselves

---

137. Ibid., 23.

138. Ibid.

139. Cf. ibid., 33.

140. Hösle, "Ontologie und Ethik," 123.

141. Ibid.

as efficient causes because we act for final causes. Once we eliminate final causality, efficient causality, too, will soon become unintelligible.

COULD FINAL CAUSALITY BE EXCLUSIVE TO HUMANS?

Very well then, we may say. Let us grant that our idea of efficient causality truly derives from the experience of ourselves as bodily agents, and let us also grant that *we* human beings act as efficient causes because we act for ends. There is thus in the world efficient causality, which we may duly ascribe not only to us but to all the world around us, and there is final causality. But why could we not restrict the latter to humans? A world in which only humans act for ends and where animals, plants, and inanimate objects do not have any purposes is one which is much better suited to scientific investigation than an alternative one where purpose pervades at least the world of the living. As Jonas points out, the scientific method's distinctive feature is analysis. To explain any entity, one has to dissect it and boil it down to its smallest parts. As we have said above, to know a thing is to know what it is made up of. Thus, the complex is explained by means of the simple, the whole by means of its parts or simplest dynamic factors.[142] In Jonas' words, "These factors are framed in such identical quantitative terms as can be entered, combined, and transformed in equations. The analytical method thus implies a primary *ontological reduction* of nature, and this precedes mathematics or other symbolism in its application to nature."[143] Put differently, the elimination of ends from any truly scientific consideration of organisms other than the human being has the advantage of making these accessible to the scientific method of analysis, which tries to explain the whole in terms of its smallest parts and, ultimately, the living in terms of the dead. In fact, "in this climate of a universal ontology of death," the attempt to integrate life into the general law of panmechanism, ultimately means to "negate it by making it one of the possible variants of the lifeless," which ironically has become the very purpose of biological science.[144] To give a simple example, instead of saying that the cat chases the mouse because the cat is hungry and hunts for food, one would more scientifically describe the matter in terms of a state of metabolic deficiency which meets with

142. Cf. Jonas, *Phenomenon of Life*, 201.

143. Ibid., 200.

144. Ibid., 11.

certain sensory data that in turn trigger a given movement. It seems that all subjective concepts can be translated into more "scientific" ones to describe the process. As Jonas comments, "In the purported 'true' description no psychological expressions (subjective concepts) appeared: homeostatic gradient, not hunger; sensory stimulus, not seeing; ceasing of tension, not satisfaction."[145] From this perspective, then, subjectivity or the character of "being-in-itself" could be predicated only of human beings but of no other being below the human person.

In this context, we may concur with Robert Spaemann when he points out that the recognition of subjectivity in another being is always a free act, an act that is based on "moral evidence" but does not derive from "theoretical compulsion."[146] He puts it this way, "The perception of the living *as* living is, as Kant first showed us, an act of freedom. . . . I can at any time decide to view a living creature which I encounter as a mere machine."[147] At another place Spaemann traces the possibility of *doubting* the inwardness of living beings—and thus the necessity of affirming their subjectivity in a free act—to the very nature of inwardness: *as* inwardness it cannot be perceived by others. To perceive another living being's pain would be to have that pain.[148] We can only perceive more or less helpful evidence for the other being's subjectivity, but never in a way that would be beyond doubt, at least in a logical sense. This is true even for the recognition of another person. For Spaemann, Nietzsche has convincingly argued the possibility of *physically* or *logically* doubting everything, but from this "it does not follow that it would be good to do that."[149] Rather, what Nietzsche has shown us is that "the necessity of positing the reality

---

145. Jonas, *Imperative of Responsibility*, 62.

146. Spaemann, *Happiness and Benevolence*, 100. For the moment of free recognition involved in interpreting the behavior of living beings as teleological, see also Löw, "Zur Wiederbegründung der organischen Naturphilosophie," 76.

147. Spaemann, *Happiness and Benevolence*, 100.

148. Cf. Spaemann, *Persons*, 180–81: "Drive, together with the pleasure and pain that accompany it, constructs an inner room without a window. No *other* creature can register the pain or pleasure felt by *this* creature. . . . That is how the Cartesians came to deny that non-human animals experience pain; for though animal behaviour presses the analogy of experience on us, we cannot be made to accept it. Inwardness, which living creatures have in common with human beings, and human beings in common with living creatures, makes it possible for humans to extract themselves from this community and treat non-human animals as mere objects. To recognize being-in-itself is never anything less than an act of freedom."

149. Spaemann, *Happiness and Benevolence*, 100.

of the living . . . is itself a kind of moral evidence."[150] There is a *moral* impossibility of doubting the reality of the other: "Whoever loves a human, whoever has friends, cannot at the same time doubt the beloved's or the friend's existence. . . . I . . . mean . . . a moral and *therefore* absolute impossibility."[151] Whether we are speaking of other human beings or of all living beings in general, for Spaemann "the affirmation of the reality of being a self is a free act. It is identical with benevolence."[152]

We have to be very clear about what Spaemann is proposing here. He does not deny the evidence for inwardness in other organisms, but only points out that our act of recognizing and affirming this inwardness cannot be forced on us by a logically stringent proof, for the simple reason that solipsism does not involve a logical contradiction. What we need to do, what we are indeed morally obliged to do, is to open our eyes to the reality of the living being, allowing the other to become real to us. Spaemann speaks of the "evidence of a perception," namely the "perception of the reality of the other and even of one's own self," which is "the basis of all moral decisions."[153]

What then is the evidence to which we have to open ourselves? Here we can come back to Jonas, who, referring to the subjectivity and goal-orientedness of animals, points out that it is much more counter-intuitive to deny it than to affirm it the moment we simply witness what animals are doing. He says, "no witness of the expressive intensity of animal acting and suffering can possibly deny" the presence of the subjective. In order to deny it, one indeed has to summon "dogmatic violence," which "nevertheless has not been lacking in the history of theory." Whatever "artificial rationale" has been given to denying inwardness to animals, the rational observer is not bound by it, "and every owner of a dog can laugh it off."[154] The mystery of animal life is truly profound, and Descartes' ma-

---

150. Ibid.

151. Ibid.

152. Ibid.

153. Ibid., 99.

154. Jonas, *Imperative of Responsibility*, 62.

Cf. also the insightful reflections by Kass, *Toward a More Natural Science*, 255–56, who argues that it is impossible to reduce organisms to simple self-producing machines, since they evidence purposes that go way beyond mere self-preservation: "A mature organism shows itself as a whole, maintains itself as a whole, and functions as a whole in characteristic ways *above and beyond merely maintaining itself.* The mockingbird delights in its own imitative sounds, the adult cardinal hops from ground to low branch and squawks for the young bird to follow suit, a coyote howls at the moon,

chine theory of animals or variants of it simply do not do justice to what is going on in a dolphin who before human witnesses leads two whales stranded on a sandbank back out into the open sea, after they had been vainly set free by human helpers a number of times but kept losing their way so as to strand again.[155] Can one interpret a border collie who has learned a vocabulary of 200 words and basically without failure goes to fetch the precise item at his owner's command as a simple automaton?[156] But even apart from such feats, every pet owner will have stories to tell about his or her animals' inwardness manifested by their capacity to learn and to respond.[157]

At this point we may want to make a 180 degree turn and look at Jonas' critique of Descartes' animal machine from the other side. Does Jonas still have something to say to us today, or is this not a point that has been brought home for a long time? Do Jonas' concerns not show that he wrote his works many decades ago? Indeed, few scientists today would want to deny that animals feel pain or that they are, to differing degrees, intelligent. If in some countries there is legislation in place that affords more protection to certain animals than to human beings in the womb, then we may wonder whether by now the pendulum has not gone to the opposite extreme. Cruelty to animals is a crime in most countries, and this is to be welcomed, though the recent debate in Spain and Austria on extending human rights to monkeys does seem to take it a bit too far,[158] always keeping in mind how little legal protection is generally enjoyed by human beings in the womb. At the end of the day, "animal rights" will always turn out to be anti-human rights, that is, they will most certainly backfire and have negative impacts on the legal protection of human beings. This is so simply because animals are not persons, which is a fact that no state legislation can change. By granting rights to non-persons, legislative assemblies implicitly say that rights are not inherent and "inalienable," based on the very nature of the beings concerned and simply

---

... the penguin struts and parades, the peacock shows off its plumage. These looks and ways serve to define the animal; its activities, taken together, are most of all what the animal *is* and what the animal *is for*."

155. Cf. *The London Times*, March 13, 2008.

156. Cf. Kaminski et al., "Word Learning," 1682–83.

157. At home, our cats, acute observers of human behavior, had learned how to open doors by jumping on the door handle. This was also the first trick our dog learned from our cats.

158. Cf. Cohen, "Unalienable Rights of Chimps," 11.

acknowledged by legislation, but rather that they are arbitrarily *conferred* by legislation as it seems fit, possibly on anyone or anything whatsoever. By the same token, legislation can then of course also arbitrarily revoke those rights. Given these dubious legal initiatives, and given that a philosopher like Peter Singer, basing a being's value on its capacity to value itself, puts a newborn baby on a lower level than "a pig, a dog, or a chimpanzee,"[159] we get the impression that what is more important today than arguing for animal inwardness is to show the *qualitative* difference between animals and human beings. This is certainly true.

And yet, we do not need to consider Jonas' critique of Descartes' animal automata as dated. His main point of critique aims at the scientific method which by its own inner logic of analysis and recomposition needs to treat the living as if it were lifeless. Here, not much has changed, and Jonas' "ontology of death" is still very much with us today. Despite some improvements regarding the legislation on keeping or slaughtering animals, including the prohibition against inflicting unnecessary suffering and pain on them, it is hard to see how the industrial keeping and breeding of cattle or poultry takes in any way seriously the inwardness or inner teleology of the animals thus kept.[160] The same of course counts for

159. Singer, *Practical Ethics*, 169–71: "I have argued that the life of a fetus (and even more plainly, of an embryo) is of no greater value than the life of a nonhuman animal at a similar level of rationality, self-consciousness, awareness, capacity to feel, etc., and that since no fetus is a person no fetus has the same claim to life as a person. Now it must be admitted that these arguments apply to the newborn baby as much as to the fetus. . . . If the fetus does not have the same claim to life as a person, it appears that the newborn baby does not either, and the life of a newborn baby is of less value to it than the life of a pig, a dog, or a chimpanzee is to the nonhuman animal."
As outrageous as Singer's position is, one needs to grant him that he is consistent. If the sole criterion for the value of a being is found in its capacity for self-awareness and its ability to value itself, and if the evil in killing a living being lies solely in causing unnecessary suffering, then his position follows: it is then not a crime to kill an embryo, because an embryo does not feel pain, nor could it then be a crime, however, to gas an adult human being in his sleep, since he or she would not so much as notice his death. (I am hesitant in advancing this point against Singer, because I am afraid that he would agree.) The alternative to Singer's position is of course that the actual presence of self-awareness or the present actualization of one's capacity for feeling pain are not the sole criteria for rights, value, and dignity. The active potency for these will be enough—and indeed already an embryo has it. In other words, one has to look at what *kind* of being the being in question is.

160. That such "unnatural" treatment of livestock can have very harmful results is demonstrated by the BSE crisis at the turn of the century. Today, it is widely acknowledged that the "mad cow disease" came about largely as the consequence of feeding cows, which are herbivores, with meat and bone meal, essentially treating them like

certain research done on human and non-human organisms, the whole point of which is precisely *not* to take seriously the being's inner teleology but to alter it and to force our own ideas on it, an attitude that is shown in many types of practices, including a recent "research" that has crossed the human-animal species line.[161] In other words, despite a raised level of awareness regarding the subjectivity of animals and increasing legislation passed on their behalf, in many ways we are still practical Cartesians.

Insofar as Cartesianism as a theoretical position on animals is concerned, already forty years ago Jonas was quite aware that even among natural scientists it was regarded more and more an untenable stance. It received a lethal blow by the theory of evolution, which "abolished the special position of man which had warranted the Cartesian treatment of all the remainder." Jonas goes on explaining: "The *continuity* of descent now established between man and the animal world made it impossible any longer to regard his mind, and mental phenomena as such, as the abrupt ingression of an ontologically foreign principle at just this point of the total flow. With the last citadel of dualism there also fell the isolation of man, and his own evidence became available again for the interpretation of that to which he belongs."[162] If there is a continuity of descent among the forms of life, then human beings are no longer the ontological oddity they were once considered to be. If we are relatives of the animals, then our experience of ourselves as organisms can be brought in as evidence for the goings-on in animals, certainly with all due respect for proportion and analogy, but evidence it can be nonetheless: "For if it was no longer possible to regard his mind as discontinuous with prehuman biological history, then by the same token no excuse was left for denying mind, in proportionate degrees, to the closer or remoter ancestral forms, and hence to any level of animality."[163]

Here indeed is the heart of Jonas' argument for the teleology inherent in the living. We ourselves are organisms. As organisms, for instance,

---

carnivores and, ultimately, forcing them to become cannibals. For this, see for instance the article by Nathanson, et al., "Bovine Spongiform", 959–69.

161. As example one can adduce the creation of hybrid beings, so-called chimera. According to an article in *The London Times*, April 2, 2008, 2, which was based on the reports of the BBC, a research team at the University of Newcastle has actually produced a hybrid with 99.9% of human DNA and 0.1% of the DNA of a cow. The chimera was created by adding human DNA to a denucleated cow egg. It lived for three days.

162. Jonas, *Phenomenon of Life*, 57.

163. Ibid.

we know what it means to see or to feel hunger. It makes all the sense in the world to understand an animal's seeing or feeling hunger in analogy to our way of doing so. Ultimately, Jonas' whole argument for the purposiveness of living beings is built "on the strength of the immediate testimony of our bodies."[164] We can know what it means to be an organism because we are organisms ourselves. On the witness of our own bodies, we know that the eye has a purpose.[165] Thus, "however complete the physicochemical analysis of the composition of the eye and of the processes attending its stimulation may be, no account of its construction and functioning is meaningful without relating it to seeing. And what is plain in so highly specialized a case is true in principle for the whole class of material things we call organisms."[166] When we speak of animal behavior in teleological terms, we are not simply opting for an alternative mode of referring to it but intend to describe "the external manifestation of the inwardness of substance." This we can do "on the evidence of each one's own organic awareness."[167]

Jonas is quite aware of the fact that Darwinism never intended to deal the final blow to Cartesianism or to vindicate the inwardness of organic life. The ironic fact that it did so, despite initially setting out as the final triumph of materialism, marks it, for Jonas, as "a thoroughly dialectical event,"[168] which, by placing the animals again in the company of the human person, restored dignity to the realm of life. We read, "Thus evolutionism undid Descartes' work more effectively than any metaphysical critique had managed to do. In the hue and cry over the indignity done to man's metaphysical status in the doctrine of his animal descent, it was overlooked that by the same token some dignity had been restored to the realm of life as a whole."[169]

Incidentally, as Leon Kass points out in this context, the benefit here goes both ways. If the animals' worth is restored by being placed into the lineage of human beings, at the same token, the status of the latter is improved insofar as now they are no longer this odd exception in a foreign

---

164. Ibid., 79.

165. In fact, Jonas goes so far to say that only a living being can know what it means to be a living being, for which reason life would actually escape the grasp of a God who is a pure mathematician: "The observer of life must be prepared by life" (ibid., 82).

166. Ibid., 90.

167. Ibid., 91.

168. Ibid., 58.

169. Ibid., 57.

world, but come to be at home again: "To be akin to the animals is to belong to the family of life. Though man remains distinct, because of self-consciousness and all that that entails, he is not alone—is not an alien freak in the mindless cosmic story. Man's metaphysical status, paradoxically, may have improved because of his newly recognized relations."[170]

Ultimately, however, all that Darwinism really shows is that the exceptional status of the human person as the only being that has inwardness or acts for ends in the realm of the living cannot be maintained. Darwinism does not decide the issue, but, by making the case for a monistic ontology, simply confronts us with the alternative either to affirm degrees of subjectivity in all living beings, including the animals, or to negate subjectivity in all these, including the human person, that is, to explain ourselves materialistically, which, however, would mean "to alienate man from himself and deny genuineness to the self-experience of life."[171]

TELEOLOGY AND FREEDOM

From what we have said above, it follows that we cannot restrict teleology solely to humans. Either we have to acknowledge it at least for the entire realm of the living, including humans and animals, or, wanting to negate it to animals and other living things, we also need to negate it to human beings. But why is teleology so important, and what does it have to do with freedom? To put the answer short: It has everything to do with freedom. Only a being that is capable of having purposes can be free. This is so because only purposiveness allows us to speak of such a thing as "a being" in the first place.[172] This can be seen by the following consideration. If there is supposed to be unity to a thing, there has to be a principle of unity, which, together with Aristotle, we can call its form. This form, however, by providing for the thing's unity, at the same time represents an end or purpose for that thing. The full realization of their form is itself an end toward which things strive. Thus, organisms strive to develop their full potential, to become all they can be. The end

---

170. Kass, "Appreciating *The Phenomenon of Life*," 5.

171. Jonas, *Phenomenon of Life*, 37.

172. Cf. Spaemann and Löw, *Natürliche Ziele*, 54: "For Aristotle *telos* is the first of all causes. If there is something which we cannot look at under the aspect of its positive or negative relation to some 'for the sake of,' then we cannot meaningfully ask for its causes, for then it is not even a determinate something" (translation my own).

of the lion is to become all that it means to be a lion. Hence, to speak of form is to speak of purpose, which is why Aristotle can say, "the 'what' and 'that for the sake of which' are one."[173] The question of purpose will decide whether something is a substance or an aggregate. If it is a substance, a truly individual thing, it will *have* purposes, because it *is* its own purpose.[174] From this it follows, of course, that only living beings are individual beings in the strict sense.[175] Thus Jonas writes, "Only those entities are individuals whose being is their own doing. . . . The realm of individuality . . . is coextensive with the biological realm as a whole."[176]

Certainly, a car, for instance, is there for a reason; it is good for something. It is nonetheless merely an aggregate, because the purposes which it servers are not its own purposes but those of its maker or owner. It is not its own purpose, and in itself it has no purpose; it is not actively striving on its own toward certain ends, nor is there any evidence for inwardness on its part. Organisms, on the contrary, and even the smallest ones, are pursuing their own ends from their inner principle, because they are their own ends. With every living thing there is at least the tendency to keep itself in being. Self-preservation is not necessarily the only nor the highest end toward which living organisms strive, but it is one that is common to all. Reproduction would seem to be another end, which higher organisms pursue sexually.[177] These ends give unity to the being which otherwise would simply be an aggregate. Having this inner unity, the organism can then affirm itself against its environment: it is a self-standing entity with a certain independence from its surroundings. This primordial demarcation of the living substance from the rest of the world, this line between the inner and the outer world, between self and not-self is the very first step of freedom. A being without purposes could

173. Aristotle, *Physics*, II, 7. Also see *On the Soul*, II, 4: "It is manifest that the soul is also the final cause of its body. For Nature, like mind, always does whatever it does for the sake of something, which something is its end. To that something corresponds in the case of animals the soul and in this it follows the order of nature."

174. Cf. also Jonas, *Imperative of Responsibility*, 98: "Every living thing is its own end which needs no further justification."

175. For a very helpful analysis of the individuality of living beings, cf. Colombo, "Vita," 169–95, particularly, 182–87.

176. Jonas, "Biological Foundations of Individuality," 233. Cf. also Spaemann, *Happiness and Benevolence*, 102, who comments on Aristotle's notion of substance: "For Aristotle the paradigm of substance is the living being."

177. Cf. Aristotle, *On the Soul*, II, 4. We will treat Jonas' lack of emphasis on this point in a later section.

not have its "own" purposes and hence could not mark itself off from its surroundings; it would be swallowed up by the outer world and have no capacity for any initiative of its own, and hence it would not be free either.

We may object that these considerations may well account for some very modest and limited animal freedom, but that freedom for humans would have to be of a different nature if it is to be real. After all, in this "biological" account, the ends, such as self-preservation or reproduction, are predetermined. They are simply given and not freely chosen. The height of freedom would seem to be in the mind, which, independent from any biological givens, can set ends for itself. Thus, we would seem to be truly free when we can set our ends for ourselves. Freedom in its authentic sense would then be found in consciousness and independent from anything that is given by the human organism. One problem with this notion of a completely undetermined freedom is that of the criteria for its choices.[178] Why does human freedom choose this rather than that? The fact is that people have the most diverse desires: there are men who want to be women; there are parents who want their perfectly healthy child to be deaf[179]; there are people with able limbs who would like to have one of them amputated.[180] Is this the height of freedom? If there is nothing pre-given, then there is no criterion by which we can say that one choice of end is better than another. The very words "good," "better," "bad," or "worse," would become meaningless because they presuppose an entity for which things are good, better, bad, or worse. What is presupposed are beings that have a certain teleology, beings that by what they are pursue certain goods that enhance their being and shun certain evils that diminish or threaten their being, in other words: things that have *natures*. If we cannot identify any goods for beings, then this also means

---

178. This problem becomes acute, for instance, when it comes to questions of genetic enhancement. Cf. Kass, *Life, Liberty*, 132: "Once genetic *enhancement* comes on the scene, standards of health, wholeness or fitness will be needed more than ever, but just then is when all pretense of standards will go out the window. 'Enhancement' is, of course, a euphemism for 'improvement,' and the idea of improvement necessarily implies a good, a better and perhaps even a best. If, however, we can no longer look to our previously unalterable human nature for a standard or norm of what is good or better, how will anyone know what constitutes an improvement?"

179. Cf. for instance, Singer's article "Shopping at the Genetic Supermarket," 143-56, who uses this example in a somewhat misleading argument, in which he attempts to sell the elimination of the patient as a commendable course of therapy for disease and in which he recommends state-involvement in questions of genetic enhancement.

180. Cf. Elliot, *Better than Well*, 208–36.

that love will become impossible if by it we mean "willing the good for the beloved."[181] If freedom is freedom from nature, we will not be able to say that, all things being equal, it is better to live than to die, to be able to hear than to be deaf, to have all one's limbs in working order than to have some amputated.

In the absence of any given ends, it would not be possible to give a rational justification of one's choices, because one has abolished all criteria that would allow one to do so. To account for one's choices, then, two ways remain: The first way is to say that they are arbitrary or ultimately absurd. What matters, then, is not *what* is chosen, but *that* it is chosen. The more emphatically and authentically I choose, the better it is. To Jonas' mind, this is what Heidegger proposes: "In the last analysis, it was a very abstract mortality that we were meant to contemplate and that was meant to make us recognize the gravity of existence. By ignoring the concrete basis of *ethics*, Heidegger's interpretation of inwardness denied itself an important means of access to this field; with this lack, ethics for him remained empty of real content. It was crucial for human beings to choose, but *what* choices they should make was not stated."[182] This is the way of modern existentialism, which, however, tacitly re-introduces at least one thing that is *given* or that is a "good," which can serve as a criterion for the goodness of one's choices: the authenticity or resoluteness of one's choosing, the affirmation of the pure will in its act of willing. The will delights in its own exercise and the abundance of its own power.[183] *What* it wills is not important, just *that* it wills. That there is no criterion for what it chooses is acknowledged as absurdity and affirmed as such.[184]

181. Thomas Aquinas, *Summa Contra Gentiles*, Book III, Chapter 90, 6: "Love consists especially in this, that the lover wills the good for his loved one." Cf. Spaemann, *Happiness and Benevolence*, 92: "Benevolence . . . presupposes the teleological structure of a living being for whom something matters, for whom something is good, conducive, or beneficial. For only then can one be benevolent to a being." For the importance of the truth about the good for the dynamism of love, also see Melina, *Sharing in Christ's Virtues*, 74–75.

182. Jonas, "Philosophy at the End of the Century," 821.

183. Cf. Nietzsche, *Genealogy of Morals*, 136, paragraph 28: "Man would rather will *nothingness* than *not* will at all." For a very original analysis of Nietzsche's thought on the will, see Arendt, *Life of the Mind*, 158–72. The section is called "Nietzsche's Repudiation of the Will."

184. What counts for Nietzsche, for instance, is precisely the will to affirm life, even in the face of meaninglessness. This capacity is the characteristic of the strong: "It is a measure of the degree of strength of will to what extent one can do without meaning in things, to what extent one can endure to live in a meaningless world" (Nietzsche,

Still, even on the height of the will willing itself to will, for the great French existentialist Jean-Paul Sartre, some nausea remains,[185] while for his compatriot Albert Camus the biggest problem of philosophy is the question suicide.[186] Nihilism, then, is the ultimate consequence: "Present day existentialism['s] . . . encounter with 'nothingness' springs from the denial of 'essence' which blocked the recourse to an idea of 'nature' of man, once offered in his classical definition by reason (*homo animal rationale*) or in the Biblical one by creation in the image of God."[187]

The other way of accounting for "freedom's" choices that are left without rational criteria to motivate them is to explain them in causal terms and hence to surrender the very idea of freedom. In absence of any goods that are there for me to pursue, in absence of any criteria that could guide my choices and motivate my acts, the most plausible thing indeed would seem to assume that my choices are *caused* by whatever chemical processes are going on in my brain or by the influences of my environment. Why I prefer to have two arms and not just one is explicable in terms of neuronal processes and chemical balances. Someone else, in whom these processes go on in a different way may find himself to have other preferences. But where there is either sheer absurdity or total determination, there can be no freedom. What we have tried to argue is that for freedom to be possible, at least some ends need to be given, which then serve as parameters in which freedom can be meaningfully exercised. The being for whom life is a good, for instance, can, on the basis of its prior acceptance of this good as a given, freely and creatively promote, cultivate and defend its life or gratuitously offer it in an act of extreme love. It is clear that such a being is much freer than a being for whom there is nothing good, that has nothing to promote, cultivate or

---

Will to Power, 318, aphorism 585).

185. Cf. Jean-Paul Sartre's existentialist novel *Nausea*. See also Jonas' comment on Sartre in Jonas, *Phenomenon of Life*, 225–26, where Jonas discusses what he perceives to be similarities between Gnosticism and existentialism: "The antinomian argument of the Gnostics is as simple as, for instance, that of Sartre. Since the transcendent is silent, Sartre argues, since 'there is no sign in the world,' man, the 'abandoned' and left-to-himself, reclaims his freedom, or rather, cannot help taking it upon himself: he 'is' that freedom, man being 'nothing but his own project,' and 'all is permitted to him.' That this freedom is of a desperate kind, and, as a compassless task, inspires dread rather than exultation, is a different matter."

186. Cf. Camus, *Myth of Sisyphus*, 3: "There is but one truly serious philosophical problem, and that is suicide."

187. Jonas, *Phenomenon of Life*, 47.

defend, that has nothing to give and is ultimately dead. Concern or interest, for Jonas, is precisely the characteristic mark of life.[188] Living beings
are concerned beings that take interest in certain things, whether it be the
cat that takes interest in the mouse or the philosopher that takes interest
in a lively discussion.

An Appreciation of *The Phenomenon of Life*

We have said that Jonas' most promising hint toward a solution of the
mind-body problem consists precisely in taking life as the fundamental
ontological category, and we have analyzed the main characteristics that
life has for him: living things are metabolizing organisms that have their
own inner teleology, i.e., they act for ends and their existence is concerned existence. Insofar as organisms are metabolizing and concerned
beings, they are marked by a certain kind of inwardness by which they
are separated from their environment. They are individual beings for
which there is the realm of self and the realm of not-self. With the higher
animals, this realm of the self, their inner perspective, seems to take on
ever higher degrees. From all the evidence we have, at least the higher
animals are susceptible to pleasure and pain. When seeing a dog leap
and dance at the return of his master after an extended absence one may
even be tempted to attribute to it the capacity to experience some of the
more spiritual emotions such as joy. It may very well be that we do greater
justice to human consciousness if we understand it as a specific kind of
that inwardness which is proper to all living beings rather than oppose
it to life. Consciousness is then life coming to its fullest awareness and
not a ghost in a machine.[189] For Jonas, then, the freedom proper to consciousness is understood as the freedom proper to the organism having
come to its very height and not as a freedom from the organism.[190] As we

188. Cf. for instance, ibid., 4, and Jonas, "Burden and Blessing," 35.

189. Cf. Jonas, *Phenomenon of Life*, 1: "The organic even in its lowest forms prefig-
ures mind, and . . . mind even on its highest reaches remains part of the organic." See
also: ibid., 57–58: "If it was no longer possible to regard his mind discontinuous with
prehuman biological history, then by the same token no excuse was left for denying
mind, in proportionate degrees to the closer or remoter ancestral forms, and hence to
any level of animality: common-sense evidence was reinstated through the sophistica-
tion of theory—against its own spirit, to be sure. . . . Inwardness is coextensive with
life."

190. Vogel in "Natural Law Judaism," 34, aptly sums up Jonas' position in this way:
"The emergence of the human mind does not mark a great divide within nature, but

have argued, a freedom of consciousness detached from organic freedom abolishes itself in its very idea. If matter is dead and consciousness is the only thing "alive," consciousness will confront matter as something completely alien and the problem of how mind and matter interact will be unsolvable. If however, consciousness is seen to grow out of the inwardness proper to living beings as such, consciousness as a heightened degree of being alive,[191] it is not something alien to the living organism, and the interaction between the organism and consciousness is a problem that is at least not entirely unsolvable since there are no two *substances* in the first place.

Maintaining the Specific Difference
between Humans and Animals

Some clarifications may be in order here. First, by emphasizing the continuity between life and consciousness, no one—neither Jonas nor anyone following him here—needs to deny the specific difference between humans and all other living beings.[192] The higher we climb on the scale of living beings, the greater the freedom of these beings from their surroundings and the greater their individuality and their degree of inwardness. From the amoebae to plants to animals, life takes on new qualities, such as sensitivity or emotions found in the higher forms of life but not

---

elaborates what is prefigured throughout the life-world."

191. Cf. also Melina, "Vita," 1520: "Life is not an object on which to perform research but the basis of all activity. Hence it is clear that the highest form of life is the one given to consciousness, with the consciousness of man. Consciousness and life cannot be opposed as subject and object. Consciousness is the most perfect level of being alive: *Quis non intellegit non habet perfecta vita* (*Summa Theologica*, I, q. 18, a. 3)" (translation my own). We will return to this question toward the end of the present chapter.

192. In fact, Jonas dedicates the chapter "Image-making and the Freedom of Man" to "determining man's 'specific difference' in the animal kingdom" (Jonas, *Phenomenon of Life*, 157) which he locates, as already the title suggests, in the human capacity to make images.

Cf. Hösle, "Ontologie und Ethik," 110: "In Jonas' philosophy, anthropology is part of the philosophy of life, for human beings are organisms. . . . Of course, this does not imply that there is no *differentiae specificae* of the human person. One difference in which Jonas is especially interested is the human capacity to create images" (translation my own).

in the lower.[193] And yet these are expressions of life and not simply something extraneous to life.

Rationality as well, with its possibility of self-awareness can be understood as a new quality that can always be seen in continuity with the other expressions that life takes on. Nonetheless, Spaemann and Löw seem right when in their critique of evolutionism they include the question about the rise of consciousness among the four occurrences of something truly new, which the theory of evolution has a certain difficulty in explaining. These four instances of the new are: 1) the big bang, 2) the transition from non-life to life, 3) the rise of consciousness out of the non-conscious, and 4) the rise of the moral out of the non-moral.[194] To see consciousness or mind in continuity with life is not to deny that its appearance is something truly new that requires an explanation as fundamental as that required for the being of anything in the first place ("big-bang") or for the coming to be of life. Even philosophically speaking these transitions—and the difficulties encountered in accounting for them—may very well speak in favor of a divine Creator. All this being granted, to try to understand consciousness, or perhaps better rationality and self-awareness, in terms of life, with whatever difficulties may flow from it, is always better than to take consciousness as an extraneous reality that is added to what is at best an animal body and at worst a simple machine.

The Possibility of the Soul's Survival after Death

Another disclaimer regards the possibility of the human soul's survival after death. This question has not been prejudged by Jonas' account. His great merit is to show that materialism and idealism are not the only alternatives to substance dualism. In fact, all three are untenable. The way out of the dilemma for Jonas is to look at life. Life shows that the alternatives substance dualism, materialism, or idealism are insufficient, since neither can make sense of life. A living being is the quintessence of what it means to be a substance. Hence, it cannot be composed of two substances; in fact, life shows that substance dualism works with a rather questionable notion of substance. Nor does the living thing fall under matter, because matter by definition is dead. Nor is it thought. Hence,

193. Cf. Jonas, *Phenomenon of Life*, 99–107.
194. Cf. Spaemann and Löw, *Natürliche Ziele*, 191.

a paradigm shift is needed that makes sense of life: an ontology of life that begins with our incontestable experience of life, looking at reality beginning with life as the fundamental category, rather than attempting to reduce all being to what is dead.[195]

As Jonas demonstrates, however, life is itself a composite reality. Living things are composites of matter and form; they are "informed" or "ensouled" matter. It is in this composite nature that the ancients and medievals saw the philosophical possibility of the soul's survival after death, at least in the case of human beings. Already Aristotle considers this at least in some ways feasible.[196] Thomas takes up the Philosopher's argument that whatever operates on its own can also exist on its own, and then strongly affirms what the latter considers a possibility, namely that there is at least one operation that the soul performs on its own, i.e., understanding, and that therefore the soul can survive the separation from its body.[197] At the same time, the Angelic Doctor is far from being a substance dualist. For him, the soul separated from its body cannot be properly called a substance, so that its mode of being is somewhat ontologically awkward.[198] Here then the significance of the Christian

195. Cf. Russo, *La biologia filosofica di Hans Jonas*, 43: "In short, what Jonas tries to do is to interpret matter starting from life rather than life beginning from matter; he would like to understand the history of being beginning from its results, as symbols of the potentialities that were at its origins, rather than understand these results as epiphenomena of elementary structures" (translation my own).

196. Cf. Aristotle, *On the Soul*, I, 1 and III, 5.

197. Cf. Thomas Aquinas, *Commentary on Aristotle's De Anima*, I, Lecture 2: "There is one activity which only depends on the body to provide its object, not its instrument; for understanding is not accomplished with a bodily organ, though it does bear on a bodily object; because . . . the phantasms in the imagination are to the intellect as colors to sight: as colors provide sight with its object, so do the phantasms serve the intellect. Since then there cannot be phantasms without a body, its seems that understanding presupposes a body—not, however, as its instrument, but simply as its object."

Cf. also: Thomas Aquinas, *Summa Theologica*, I, 75, 2: "The intellectual principle which we call the mind or the intellect has an operation 'per se' apart from the body. Now only that which subsists can have an operation 'per se.' For nothing can operate but what is actual: for which reason we do not say that heat imparts heat, but that what is hot gives heat. We must conclude, therefore, that the human soul, which is called the intellect or the mind, is something incorporeal and subsistent."

198. Cf. Thomas Aquinas, *Summa Theologica*, I, 29, 1, ad. 5: "The soul is a part of the human species; and so, although it may exist in a separate state, yet since it ever retains its nature of unibility, it cannot be called an individual substance, which is the hypostasis or first substance, as neither can the hand nor any other part of man; thus neither the definition nor the name of person belongs to it."

dogma of the resurrection of the body comes in: the separated soul is not a substance; it awaits and longs to be reunited to its body.

The Ontology of Life and the Problem of Death

Within the context of the ontology of life that Jonas proposes, what sense can be made of death? Is not death the great nullifier, which, by letting the living being return to the ground whence it was taken, proves this ground and not the living substance to be the ultimate reality? If all life ends with death, does this not show that dead matter is the more fundamental reality, so that for the purpose of knowledge, it makes all the sense in the world to try to explain the living thing materially, that is, in terms of what is dead?

To this objection Jonas gives a striking answer that has two aspects, one referring to all living beings as such and one appertaining more specifically to the human person. In his essay, perhaps somewhat provocatively entitled "The Burden and Blessing of Mortality," published toward the end of his lifetime, Jonas writes, "If it is true that with metabolizing existence not-being made its appearance in the world as an alternative embodied in the existence itself, it is equally true that thereby to be first assumes an emphatic sense: intrinsically qualified by the threat of its negative it must affirm itself, and existence affirmed is existence as a concern. Being has become a task rather than a given state, a possibility ever to be realized anew in opposition to its ever-present contrary, not-being, which inevitably will engulf it in the end."[199] Non-being becomes a reality only for entities that *are* in the emphatic sense that they show concern for their being. Yet they show concern for their being, which is their life, only because their being can be taken away from them in death. Jonas puts it this way, "The basic clue is that life says yes to itself. By clinging to itself it declares that it values itself. But one clings only to what can be taken away. From the organism, which has being strictly on loan, it can be taken and will be unless from moment to moment reclaimed."[200] The reason that living things *are* emphatically, i.e., the reason that they show concern for their being and value it, is precisely because of death, because they are constantly pitted against non-being, so that, in a somewhat paradoxical way, death turns out to be one of the very conditions of the

199. Jonas, "Burden and Blessing," 35.
200. Ibid., 36.

possibility of life as concerned existence, that is, an existence that affirms and treasures itself.

All of this of course also applies to humans, though here Jonas takes his reflections a step further. Evidently he does not deny the burden of mortality and the gravity of a premature and untimely death, the death of people whose life was a story that had hardly begun and was abruptly ended on the first pages.[201] If he speaks of a "blessing" of mortality for the individual person he emphasizes that this "is true only after a completed life, in the fullness of time" and that in fact we have the duty "to combat premature death among humankind worldwide and in all its causes—hunger, diseases, war, and so on." [202] To the burden of mortality no doubt also belong the pains of separation and loss as well as the pains connected to illness, which is as such a precursor of death. But now how can Jonas speak of the "blessing" of mortality? Here, he points out that mortality belongs to the human condition no less than "natality," which is the famous term coined by his friend Hannah Arendt and which refers to the fact that human beings are born. Natality and mortality belong together; the one can only be had at the cost of the other; they are two sides of the same coin. The newness, spontaneity, and freshness that is connected to natality, the influx of new initiatives and new ideas brought in with the coming of new persons and new generations, the possibility of new beginnings as each generation faces the world in a new light—all this can only be had if the older generations gradually give way to the younger ones.[203] At times, for instance, the only solution to some political problems is a biological one. Once people have acquired certain views, made certain experiences and reached a certain age, they are very unlikely to change their positions. Relations between nations can be in a

---

201. Cf. Jonas and Scodel, "An Interview," 347–48. In the interview, Jonas is asked about his concern about his own mortality (in fact, Jonas publically stated that he had written *The Imperative of Responsibility* in German first and not in English because he did not think he would still live long enough to finish it if he had drafted it in English). Jonas responds by saying that "it is natural and proper for life to have an end. And the whole idea of going on and on and on is deeply repugnant to me. I think it flies in the face of what life is about. It is finiteness, its finitude belongs to it." However, then he immediately goes on to qualify his statement: "Of course, I have a lot against premature death. That's a different thing. And that can be very tragic and very sad. But death, I think, is described in the Bible, 'And he was assembled to his forebears old and sated with days.' 'Sated with days,' one can be sated with days."

202. Jonas, "Burden and Blessing," 40.

203. Cf. ibid., 39.

deadlock until on both sides a new generation comes in that sees things from a different perspective and comes up with new solutions.[204] In fact, for Hannah Arendt the source of novelty brought by new generations, i.e., "the fact of natality," is nothing less than "the miracle that saves the world, the realm of human affairs, from its normal, 'natural' ruin . . . . It is, in other words, the birth of new men and the new beginning, the action they are capable of by virtue of being born."[205] But this constant influx of new people, Jonas points out, can be had only at the price of the constant parting of the older generations, for which reason he argues that the general human condition of mortality can plausibly be called a blessing for humankind as such.

This consideration may well be to the point with regards to humanity as such—it is good for humanity that new generations are coming in, while the older ones are leaving—but what about the individual? Is not death the greatest evil, at least for the individual person? In his *The Imperative of Responsibility,* Jonas makes the striking statement that "perhaps a nonnegotiable limit to our expected time is necessary for each of us as the incentive to number our days and make them count."[206] In this one short sentence Jonas has managed to capture an extraordinary depth of meaning, which deserves further reflection. For this we may turn for a moment to Robert Spaemann, who, apparently following the lines of Jonas' insight, spells out its implications in greater detail, pointing out how it is precisely the awareness of our death that gives value to the

---

204. Thus, Francis Fukuyama points to "the deleterious consequences of prolonged generational succession in authoritarian regimes that have no constitutional requirements limiting tenure in office. As long as dictators like Francisco Franco, Kim Il Sung, and Fidel Castro physically survive, their societies have no way of replacing them, and all political and social change is effectively on hold until they die" (*Our Posthuman Future,* 65). He continues explaining how what holds true for politics also applies to other fields. Hence, "the discipline of economics makes progress one funeral at a time" and "the survival of a basic 'paradigm'"—in whatever scientific discipline—"depends not just on empirical evidence, as some would think, but on the physical survival of the people who created the paradigm" (Fukuyama, *Our Posthuman Future,* 66).

205. Arendt, *Human Condition,* 247.

206. Jonas, *Imperative of Responsibility,* 19. He later takes up this formulation in Jonas, "Burden and Blessing," 40. Cf. also Leon Kass' reflection on the subject, which are very much along Jonas' lines: "Could life be serious or meaningful without the limit of mortality? Is not the limit on our time the ground of our taking life seriously and living it passionately? To know and to feel that one goes around only once, and that the deadline is not out of sight, is for many people the necessary spur to the pursuit of something worthwhile" (Kass, *Life, Liberty,* 266).

present moment: "'Significance' is meaning 'toughened' by the conscious-ness of finitude—by which is understood that it asserts itself in the face of death."[207] If it were not for death, then everything we can do today we could still do tomorrow—there would really be no reason to do it today; every day would be as good or as bad as the other and nothing would have any urgency nor any uniqueness to it. Given that there will be an end to our lifespan, however, every point in the time of our life is unique.

Furthermore, in an original consideration of the grammatical future perfect tense, Spaemann argues that death is the finalizer that not only ends but in a certain way completes our lives. It is the point of the final "will-have-been": "The future perfect is the form that eternalizes things. Since everything present is also about to have been—for ever and ever—it belongs already to the dimension of the timeless. What *was* future *be-came* present; what *is* present *becomes* past; but past it *will* remain for all futurity. An event of which we had to say when it was present that it would cease one day *to have been,* would be unreal, even while it was present."[208] This means that everything I do, every minute little action I perform in the time of my life has eternal significance because forever it will have been. Precisely *because* my life is finite, every action I place in time is irrevocable. As the action moves into the past, it becomes fixed and eternally unchangeable as the past itself. Every movie I watch, every word I say—forever it will have been, forever it will have taken its place in the story of my life and in the story of the world. Besides, not only did I do this, there are also a million of other things I did not do because I did this; everything I do precludes all other options which forever will remain unfulfilled as options that were open at that specific time.

Spaemann gives the example of enjoying a good glass of wine with an old friend in a beautiful country setting. Does not the certainty of death, the thought of ultimate separation nullify the preciousness of the moment and reveal it to be nothing but a farce devoid of meaning? One could think of it in this manner, but for Spaemann there is an alterna-tive manner of looking at it. There is a way in which it is precisely death that gives this moment its preciousness and its significance. It is precious because it is unique; it can be now and only now, and it will have eternal significance as now. One will always be able to say, "it is good that it occurred":

207. Spaemann, *Persons*, 119.
208. Ibid., 121.

Such a feeling would not be threatened by the imminent end of life and of the meanings that derive from life, but would actually be awakened by it. . . . It means, "It is, and will remain, good that this fleeting moment occurred and that its significance is unveiled." Meaning, together with the feeling it engenders, is pulled out of the contingent and relocated in the timelessness of "significance." If the friends savour their food and wine, it is not because instinctual needs are satisfied, but because the whole complex of need and satisfaction has been lifted out of its relativity. . . . The event in its totality . . . appears a something of which it can be said—both now and always—that it is good that it occurred. The nothingness of what succumbs to time is converted into preciousness.[209]

Let us then return to Jonas. To show the doubtful desirability of earthly immortality, Jonas refers to the literary character of the Struldbrugs that appear in Jonathan Swift's novel *Gulliver's Travels,* and who are in fact immortal. On hearing about them, "Gulliver is enraptured by the thought of their good fortune and that of a society harboring such fonts of experience and wisdom. But he learns that theirs is a miserable lot, universally pitied and despised; their unending lives turn into ever more worthless burdens to them and the mortals around them."[210] These immortals do not live the admirable life of the gods. They have separated themselves from the mortals; the social interaction among themselves is limited; there is no marriage.[211] Jonas points out how in the novel the immortals can remember only what they have learned in their youth.[212]

209. Ibid., 119–20.

210. Jonas, "Burden and Blessing," 39.

211. While Gulliver's account, related here by Jonas, links the end of marriage simply to the fact that the immortals dislike each other's company, so that "'those who are condemned without any fault of their own to a perpetual continuance in the world should not have their misery doubled by the load of a wife'—or a husband," as Jonas is quick to add (Jonas, "Burden and Blessing," 39), Leon Kass points out the intrinsic relation between marriage—or at least sexuality—and mortality: "In sex, life is *not* just self-centered individuality, on the contrary, sexual desire, in its deepest meaning, is self-sacrificing. . . . To put it sharply: any animal engaged in sex is, willy-nilly and albeit only tacitly, voting with its genitalia for its own disappearance" (Kass, "Appreciating *The Phenomenon of Life*," 11). Cf. also Kass, *Life, Liberty,* 273: "Children and their education, not growth hormone and perpetual organ replacement, are life's—and wisdom's—answer to mortality." But on a different occasion Jonas, too, argues along these lines: "If we abolish death, we must abolish procreation as well, for the latter is life's answer to the former" (Jonas, *Imperative of Responsibility,* 19).

212. Cf. Jonas, "Burden and Blessing," 40.

For our author this is a very feasible scenario and can be traced back to the fact that even if our bodies were to perdure, our minds would still be finite. According to him, we can only "store" so much memories and information, and the only solution to this problem would be to periodically "delete" our memories and make room for new ones. With that, however, we would lose our personal identity.[213]

Always keeping in mind that we are dealing with fictional characters here and not with actually observed phenomena, we may still fail to be entirely convinced by Jonas' explanation for why the immortals cannot remember anything except what they have learned in their youth, as the mind is understood here somewhat reductively in analogy to a computer hard-drive that at some point is simply "full."[214] But we may wonder whether there may not be yet another account for the immortals' inability to remember, even if we admit the mind's potential openness toward the infinite. As Charles Péguy points out, there is something remarkable about the new, about what we do for the first time, about the "first day":

> All that begins has a character that is never again recovered.
> A strength, a novelty, a freshness like dawn.
> A youth, an ardor.
> A spirit.
> A naivcté.
> A birth that is never again recovered.
> The first day is the most beautiful day.
> The first day is perhaps the only beautiful day.[215]

Our spirits are kept alive by new discoveries, by hopes yet unfulfilled but worth striving for, by experiences had for the first time. Now it is the fate of the earthly immortals that they no longer have the experience of the new. Hence, they start to lose all interest in anything and their desires are dulled. Though their bodies live on, the youthful vigor of their spirits declines to the same degree in which they grow in experience. As they grow older, they have seen everything that there is to be seen; they had their hopes fulfilled or disappointed and in either way have become more realistic as to what to expect from life; there is nothing left to be

---

213. Cf. ibid., 40.

214. Cf. ibid.: "There is a finite space for all this, and those magicians would also periodically have to clear the mind (like a computer memory) of its old contents to make place for the new."

215. Péguy, *Portal of the Mystery of Hope*, 20.

done for the first time.[216] Everything becomes a repetition as life grows duller and duller. Ultimately life may become so boring that there is not even enough excitation to make any impression on our memory—we do not tend to remember unimportant, boring, or routine events. We do not remember whether we have locked the car, though we have just walked away from it a minute ago. We do not remember because we have locked it a thousand times before, and because as such it is an insignificant routine event. For the immortals of *Gulliver's Travels*, their every activity must be just like that.

Returning from the discussion of a fiction to a reflection on our concrete lived reality, we can say that of course one joy remains as we get older, even if we have seen it all before, namely to watch our children and our children's children take our place, follow in our footsteps or find their own ways.[217] But for this it is necessary that we ourselves slowly decrease and withdraw to make room for them. On the hypothesis of immortality, we would hardly retreat voluntarily from our positions and offices, so that the younger ones would have to expel us by force. The greatest joy proper to old age—to see our children prosper—would thus be turned into the bitterness of a struggle for influence.

With such a bleak picture painted of the thought of earthly immortality, we may wonder whether the immortality after death proclaimed by Christianity is such a promising thing after all. Here we must remember that for the Christian tradition the immortal existence in heaven is not thought of in analogy to the present life as an unending sequence of events, one thing forever following upon another. This could hardly be appealing, and Adam and Eve's expulsion from Paradise, lest they eat of the tree of life and live forever (cf. Gen 3:22), may in fact be seen as an act of God's mercy toward fallen humankind rather than an act of vindictive punishment. Christian theology rather understands heaven as

216. For this consideration of the effects of aging on the mind, also see Kass, *Life, Liberty,* 272. On the vigor of youthfulness and on "the unique privilege of seeing the world for the first time and with new eyes," also see: Jonas, *Imperative of Responsibility,* 19.

217. One may wonder what is the greater success for a politician like Helmut Kohl, for instance, to manage to stay in power as the head of the German government for 16 years or to see his own political daughter Angela Merkel—the woman he deliberately built up—become chancellor seven years after him. Although the personal joy at this achievement may be much diminished by her lack of loyalty toward him during the time of his crisis at the turn of the millennium, it is his objective success nonetheless.

an eternal now,[218] as Augustine's famous standing present,[219] which, with all the difficulties we have in imagining this new mode of existence, will most certainly not be dull or boring.

In sum, death does not annihilate purpose in Jonas' ontology of life.[220] Rather mortality turns out to be that part of the human condition that makes purpose possible, as only in confrontation with its possible non-being, life becomes concerned existence, capable of taking interests and of having purposes in the first place. We are thus ultimately reminded of the thought of Jonas' teacher Martin Heidegger, and Jonas indeed suggests that "the deeper insight of Heidegger is right—that, facing our finitude, we find that we care, not only whether we exist but how we exist."[221] It is our mortality that spurs us on to live "authentically," and it is death that qualifies our existence as care or concern.

Yet there are at least two important differences between Heidegger and Jonas on the topic of mortality. Firstly, Heidegger restricts his analysis to human existence (*Dasein*), in which he sees the key to unlocking the mystery of being. Jonas, in contrast, expands his examination to include every living thing and asks what it means for it to be confronted with death. He, then, does not see the human person alone as the point of access to being, as does Heidegger, but rather for him the phenomenon of life in general is the way for any kind of metaphysics to approach its task. Secondly, for Heidegger the being-toward-death isolates the person, who only in his or her isolation achieves authentic existence. For Jonas, in contrast, because of its needfulness and, ultimately, mortality, life is

218. Cf. also Jonas, *Phenomenon of Life*, 270: "It may well be that the point of the moment, not the expanse of the flux, is our link to eternity."

219. Cf. Augustine, *Confessions*, XI, 14, 17: "But if the present were always present, and would not pass into the past, it would no longer be time, but eternity."

220. The question of course remains whether a purely philosophical account of death such as Jonas intends to give can ever be existentially satisfactory, or, in other terms, "whether true patience [in front of sickness and death] can be had without God" (Hauerwas, "Practicing Patience," 175). In this, what we consider a fundamentally philosophical essay, we will be content with simply raising the question without however pursuing it further. We think that Jonas' philosophical reflections did succeed in showing that death does not need to be the destroyer of the meaning of this earthly life, but that death in some ways helps us to treasure life. Nonetheless, how much these philosophical reflections can actually help us in concrete situations, when our life slips away in sickness or when we are confronted with the loss of loved ones, is another matter.

221. Jonas, *Phenomenon of Life*, 234.

inherently relational.[222] It does not "live to itself nor die to itself" as one could say adapting the words of the Apostle Paul (cf. Rom 14:7). Every living being is intrinsically dependent both on the animate and the inanimate environment; it is in constant contact and in constant interchange, so that even its death is a "communal" event, insofar as other beings will always be affected by it. Besides, for Jonas already the threat of death, the living thing's being-toward-death, does not at all cause it to withdraw into itself, but rather opens it up to go out of itself with the aim of seeking sustenance and maintaining itself in being.[223]

The Organism and Love

At the end of our exposition of Jonas' thought on "the phenomenon of life," there is an important question we must raise, however, about his account of the living organism. He seems to focus exclusively on the need-aspect that characterizes living things: they enter into relationship with other beings because their existence is precarious. They need sustenance and hence must enter into a process of constant interchange with their environment on the pain of ceasing to be. All this seems true enough, but there also appears to be more to living beings than this. Despite the precariousness of their existence, despite their constant struggle against death—a struggle they will ultimately lose—organisms are also profoundly marked by their fruitfulness. Inherent to organic existence there is not only the individual's struggle against non-being, caused by want, but also its generosity to pass on life to others of its kind, flowing from a superabundance. The organism is marked by both: lack and superabundance, the tendency to struggle for individual survival and the tendency generously to give of itself, transcending the individual so that others, too, can be. Nobel prize laureate Wendell Stanley goes so far as to define life, not by metabolism as does Jonas, but by its reproductive capacity: "The essence of life is the ability to reproduce."[224] As Kass points out, to such a definition one may of course object that in this case "*one* rabbit

222. Cf. ibid., 4: "Life always exhibits . . . the polarity of being and not-being, of self and world, of form and matter, of freedom and necessity. These . . . are forms of relation: life is essentially relationship."

223. Cf. ibid., 4–5: "Relation as such implies 'transcendence,' a going-beyond-itself on the part of that which entertains the relation."

224. Stanley, "Penrose Memorial Lecture," 318.

is dead,"[225] or, in Jonas' terms, that "reproduction and sociality are not *indispensable* functions of life for an individual animal *qua* living thing; a sterile or celibate being, and even the last member of his species living on a deserted island, is nonetheless very much alive."[226] However, in their responses both Jonas and Kass do not seem to distinguish between two senses of "ability" or "possibility." In one sense it means an inner capacity that awaits its necessary conditions in order to be actualized; in another sense it refers to the same capacity together with the obtaining of the necessary conditions. To put it simply, Mozart can always play the piano in the sense that he knows how to do it, although he can *actually* play the piano only when one is really available. A single rabbit on a lonely island can still reproduce in the same sense in which Mozart, confined to the same island without a piano, can still play the piano, having the capacity that awaits its necessary conditions.[227]

However that may be, one cannot deny that the aspect of procreation and fruitfulness is an essential characteristic of the living, a discussion of which is somewhat lacking in Jonas' philosophy of the organism. In a private conversation with Leon Kass about the issue, Jonas admitted that much.[228] The omission, however, was not deliberate,[229] and Jonas responded to Kass that if he "were to rewrite the book [*The Phenomenon of Life*] . . . he would make the necessary qualifications and corrections."[230] Jonas himself traced the omission back to Heidegger's influence on him, an influence that persisted even as he had come formally to disavow many aspects of his former professor's teachings.[231] For Heidegger, in any case, human existence becomes authentic in the utter isolation of *Dasein's* facing death, a mode of being which is certainly not one that would make us think of fruitfulness.

225. Cf. Kass, "Appreciating *The Phenomenon of Life*," 11.

226. Ibid. Kass is here reporting the quintessence of Jonas' response when he asked him about the issue.

227. For the Aristotelian definition of "potentiality" and a critique of the Megarics, who ultimately claim that only the actual is possible, see Spaemann and Löw, *Natürliche Ziele*, 45–46.

228. Kass, "Appreciating *The Phenomenon of Life*," 11.

229. Vittorio Hösle, in his "Ontologie und Ethik," 117, sees the reason for this oversight in an inherent weakness of Jonas' phenomenological method, which "lacks any criterion that would enable it to claim the completeness of its description" (translation my own).

230. Kass, "Appreciating *The Phenomenon of Life*," 11.

231. Cf. ibid.

In his very helpful article "Love and the Organism," José Granados, too, notices the lack of due emphasis on procreation in Jonas' thought. However, he goes beyond that issue by pointing to another important fact of organic life: symbiotic relationships. Living beings' relationships are not only marked by need and fulfillment, as a lion "transcends" itself to go after the gazelle in order to maintain itself in being, or by relationships engendered by sexuality, but, as Granados points out, there are also symbiotic relationships of cooperation, which Jonas overlooked:

> Let us now address a weakness in Jonas' approach: his analysis of the organism takes insufficient account of the interaction between different organisms. . . . As has been demonstrated by some biologists, one important factor by which organisms evolve is the symbiotic association among organisms. In the most developed living beings, we find associations in groups, like bees in a hive or members of a herd, in which each organism subordinates itself to a more complex structure. That holds true even for the most primitive forms of life; we can mention, for example, the recently discovered fact that if an amoeba encounters difficulties in the process of self-division, it "calls" another amoeba who acts as a kind of midwife to assist the first amoeba.[232]

Active cooperation among organisms is of course not restricted to members of the same species. There are many types of symbiotic relationships formed among the most diverse organisms, from which all parties involved benefit. Just to give an example, it seems that the human body is the home to close to three pounds worth of symbiotic bacteria, most of them in the digestive tract, aiding the human organism in various tasks.[233]

What follows from these latter remarks for Jonas' ontology of life and his notion of freedom? What would happen if we "added" the missing dimensions of procreation/fruitfulness and symbiosis/cooperation to Jonas' thought? Jonas' great contribution to metaphysics, but also to anthropology, lies precisely in this, to see freedom present already on the level of the organic, namely in metabolism, and to trace its growth on a continuous line along the ascending order of being, so that the freedom

---

232. Granados, "Love and the Organism," 30.

233. Wilson, *Microbial Inhabitants of Humans*, xvii: "An adult human being consists of ten times as many microbial cells as mammal cells, and he or she carries around approximately 1.25 kg of microbes. Knowing this, who could fail to be intrigued by the microbial component of that mammal-microbe symbiosis known as a 'human being.'"

proper to human beings—which for Jonas lies mostly on the level of consciousness with its ensuing capacities for self-determination and self-transcendence—does not fall from the sky but is firmly rooted in life itself. Thus he writes in one of his later essays, which may well serve as a summary of his thought on the issue, "The great contradictions that man discovers in himself—freedom and necessity, autonomy and dependence, ego and world, connectedness and isolation, creativity and mortality— are present *in nuce* in life's most primitive forms." Going up "the ascending order of organic capabilities and functions: metabolism, motility, and appetite, feeling and perception, imagination, art, and thinking" means to go up "a progressive scale of freedom and danger, reaching its pinnacle in man, who can perhaps understand his uniqueness in a new way if he no longer regards himself in metaphysical isolation."[234] Hence, human persons are at the highest point of the "scale of freedom," and the most distinguishing faculty among their "organic capabilities and functions" lies precisely in thought. For Jonas, consciousness is the highest form that life takes on, and this freedom of thought is already prefigured at the very beginnings of life, namely in metabolism.

Now basing himself on Jonas' idea that the higher phenomena of freedom are ontologically based on the types of freedom already present in the most primitive forms of life, Granados makes an intriguing suggestion, proposing that precisely the evidence of life in its early forms actually entitles us to take things up one step further. Even on the most primitive level like that of the amoeba, living beings are not only metabolizing beings aiming at their own survival, but they are always cooperating beings as well. The openness toward others is present at the first inklings of life in symbiotic relationships and is increased in the higher organisms through the arrival of the passions and sexual reproduction, ultimately finding its pinnacle in human love, where these passions find their integration.[235] It is thus in the experience of love that the highest form of self-transcendence is achieved and thus also the highest form of freedom.[236] In other words, the evidence of organic being, to which Jonas

---

234. Jonas, *Mortality and Morality*, 60.

235. Cf. Granados, "Love and the Organism," 32: "An essential element of the experience of love is the integration of the world of the passions. That is the way love makes it possible for another person to affect one's own world; this affection sets man in motion towards the achievement of a communion that must be built up in time in a free and active way by the patient integration of all the affective levels."

236. Cf. ibid.: "We see, then, that by means of this love man is able to transcend

himself appeals, suggests to us that the highest form of freedom is not located in consciousness but in the communion of persons in love. As Granados puts it, "The communion of persons, and not self-consciousness alone, becomes the final stage of this process of transcendence that began with the first manifestations of life."[237] And this freedom is indeed already prefigured on various levels of organic being, from symbiotic relationships to animal passions to human passions integrated in love.

It is important to note that this way of going beyond Jonas does not compromise one of his greatest concerns, namely the struggle against dualism. Insofar as life's transcendence in communion "is present in an analogical way at the origins of life," it will continue to be possible to avoid a dualistic understanding of the human being. "This transcendence," as Granados puts it, "is going to be rooted in the body, which allows the living being contact with other forms of life that can enrich his own world."[238]

Taking our clues from Jonas' ontology of life and basing ourselves on his claim that already the primitive organic forms in some ways prefigure the higher forms of freedom, we are thus entitled to suggest that where life comes to its fullness is not consciousness but *communion*, namely the communion of persons in love. We are in fact reminded here of the words of John's Gospel: "This is eternal life, that they know thee, the only true God, and Jesus Christ whom thou hast sent" (John 17:3 *RSV*). As Livio Melina points out, this kind of "knowing" is not to be understood in intellectualistic terms, but it rather implies a way of entering into communion.[239] Thus, for the Christian faith, the fullness of life—where life is most fully revealed and reaches its highest point—is in communion, namely in communion with the Trinitarian God who is communion himself.

---

himself and discover his true freedom via an increase of dependence and vulnerability."

237. Ibid.

238. Ibid.

239. Cf. Melina, *Epiphany of Love*, 16.

2

Jonas' Philosophy of Responsibility

NOW INSOFAR AS THE question of reproduction is concerned, it is in some ways ironic that a more detailed theoretical account of it is missing in Jonas' philosophy of the organism. Of course he was not blind to the transcendence toward others and toward the future implied by sexuality and reproduction, which for humans is the basis of familial societies. As Kass points out, "Now all of this Hans Jonas knew and, I want to say, knew in his bones. On few subjects was he more eloquent than on the wisdom of finitude and the redemptive possibility of perpetuation, a possibility, because now become precarious, he made the cornerstone of his teaching of responsibility."[1] What Kass refers to here is Jonas' *The Imperative of Responsibility,* in which precisely the parent-child relationship serves as a major paradigm for the key concept under discussion: responsibility. In this book Jonas argues for the need of a new ethics of responsibility, which is not to replace, but to supplement the more traditional approaches.[2] Much of his thought on the organism is either presupposed

---

1. Kass, "Appreciating *The Phenomenon of Life,*" 12.

2. The same Jonas who begins his work with the rather provocative claim that "all previous ethics" had tacit premises in common that "no longer hold" (Jonas, *Imperative of Responsibility*, 1), also says, "To be sure, the old prescriptions of the 'neighbor' ethics—of justice, charity, honesty, and so on—still hold in their intimate immediacy for the nearest, day-by-day sphere of human interaction" (Jonas, *Imperative of Responsibility*, 6). Jonas has no intention to *break* with previous ethics, although he believes that certain developments in technology require us to rethink some fundamental

or reiterated here, and it will certainly help the interpreter to read *The Phenomenon of Life* and *The Imperative of Responsibility* together.

Though this time the word "freedom" does not make it into the title—neither in the English nor in the German edition of Jonas' work—freedom is of course the great presupposition of the book. Insofar as responsibility, as he will define it, is the capacity to respond to a duty that issues forth from being, it presupposes freedom: I can only respond to the call that addresses me in the face of what is both precious and vulnerable when I am in possession of myself and can determine myself, i.e., only when I am the true author and initiator of my acts. As Jonas points out, responsibility presupposes power, and at the very least it would seem to presuppose that power over myself which is tantamount to the freedom of self-determination and self-transcendence. Thus when Jonas says that the goal of the new ethics that he is proposing is to ensure the possibility of the continued existence of responsible beings, then he necessarily also asks about the conditions of the possibility of freedom. In what follows we will trace the steps of his argument, in which he first presents the ambiguity inherent to some of the achievements of modern technology, then proposes the outlines of his ethics of responsibility, and finally examines which political system could be most suited, in the face of global threats, to guarantee the continued existence of human beings on earth.

Analysis of the Modern Predicament and Justification for the Need of a New Ethics

Jonas begins his book with an analysis of how modern technology is different from previous human interventions into nature, which of course have always occurred. Our author cites a famous passage from the Antigone choir: "Many the wonders but nothing more wondrous than man."[3] The song reveals great admiration for human inventiveness and power over nature: human beings tame wild beasts, they furrow the land, they brave the sea. And yet, Jonas points out how in the mind of the ancient author, human beings, with all their ingenuity, are still small in comparison to the great elemental powers of nature, which constitute a cosmic order that is forever fixed and unchangeable. That humans could directly

---

presuppositions. For his use of hyperbole as "a perfectly legitimate rhetorical device for shocking us into a new awareness," cf. Bernstein, "Rethinking Responsibility," 19.

3. Jonas, *Imperative of Responsibility*, 2, citing Sophocles, *Antigone*, line 325.

intervene into this order was unimaginable to the ancient mind. In sum, "nature was not an object of human responsibility—she taking care of herself."[4] Corresponding to this view of nature and of the cosmic order, there is a traditional ethics which for Jonas was characterized by four main points[5]:

1. It did not include human dealings with nature among its objects. The realm of *techne,* of fabrication and production, was considered ethically neutral.

2. Ethical significance was restricted to the realm of human interaction, so that one could say that traditional ethics was anthropocentric.

3. This human being, on which ethics focused, was considered unalterable and constant. Human nature was a given and not a potential object for human art and manipulation.

4. Traditional ethics was essentially a neighbor ethics, i.e., it was concerned with what lay in the proximity of action in space and time. It did not need an extended knowledge of the future to judge human action because human power was limited: "The short arm of human power did not call for a long arm of predictive knowledge."[6]

While the old precepts of traditional neighbor ethics still count, Jonas argues that given the new character of human action, with its increased power over nature, things have drastically changed for ethics and the four points above have to be revisited. As nature has become critically vulnerable to human technological intervention, "the whole biosphere of the planet has to be added to what we must be responsible for because of our power over it."[7] We can no longer limit the field of our moral responsibility to what is proximate in space and time, since technological praxis initiates irreversible and cumulative causal chains.[8] The cumulative character of technological inventions and interventions, where the only constant factor is change itself, presents a grave difficulty for our ability to evaluate and predict possible outcomes. We continually find ourselves in situations without precedence with no previous experience

---

4. Jonas, *Imperative of Responsibility,* 4.
5. Cf. ibid., 4–5.
6. Jonas, *Imperative of Responsibility,* 6.
7. Ibid., 7.
8. Cf. ibid.

to draw from.[9] Hence Jonas speaks of a chasm between our power to predict and our power to do, so that "recognition of ignorance becomes the obverse of the duty to know and thus part of the ethics that must govern the evermore necessary self-policing of our outsized might."[10]

The irreversibility of the results of our new technological action, in turn, needs to make us especially concerned with the beginnings. We need to evaluate critically whether what we are striving for is really desirable, because once it is there, it will be a condition of the possibility of future action.[11] For instance, it would have been much easier in the past to forego the invention of the nuclear bomb by directing research elsewhere, than it is today to rid ourselves of it, now that it has been invented and become part of the inventory of really existing things. Thus, on a different occasion, Jonas cites Robert Oppenheimer who said after Hiroshima, "The natural scientist has become acquainted with sin," referring to nuclear physics and the bomb.[12] In this context Jonas even raises the question of the self-censorship of science in the name of responsibility.[13]

In short, for Jonas, given the new character of human action, which now has power over nature in general and over human nature in particular, we need a new ethics, an ethics of responsibility for the future of human nature, which for the first time in our history has become vulnerable. Importantly, concern and responsibility for human nature—or "the image" as Jonas calls it at times—includes concern and responsibility for nature as a whole. Nature as the global order of Planet Earth presents the condition under which human beings exist, and its unimpaired state is thus necessary for authentic human life. Besides, we need to ask seriously whether nature does not have a moral claim on us even in its own right, given that now we have power over it.[14]

9. Cf. ibid.

10. Ibid., 8.

11. Cf. ibid., 32. For the important observation that even the things made by human hands always and necessarily turn into conditions of our existence, see also Jonas' friend and colleague Arendt, *Human Condition*, 9: "The things that owe their existence exclusively to men nevertheless constantly condition their human makers."

12. Jonas, *Technik, Medizin und Ethik*, 76, (translation my own).

13. Cf. ibid., 80.

14. Cf. Jonas, *Imperative of Responsibility*, 8.

The New Categorical Imperative and Its Foundation

Having analyzed the new conditions of human action in the technological age and having justified the need for a new ethics, Jonas immediately
continues by proposing what he calls a new categorical imperative, one
that is, admittedly, directed more at public politics than at private action. Presenting this imperative in what is still an early section of his
book, he will attempt to justify and ground it throughout the remainder
of the work. Given our new power to extinguish the human race or to
render miserable the conditions of its existence, the continued presence
of authentic human life in the world has itself become a matter of human
obligation. Thus, the new categorical imperative, namely the imperative
of responsibility, says, "Act so that the effects of your action are compatible with the permanence of genuine human life'; or expressed negatively:
'Act so that the effects of your action are not destructive of the future
possibility of such life.'"[15]

Logical Coherence?

The main question that must be raised here is that of the foundation of
this imperative. What is it based on? Jonas points out that it is very different from Kant's categorical imperative or universalizability principle,
which says, "Act so that you *can* will that the maxim of your action be
made the principle of a universal law.'"[16] The foundation of this principle,
Jonas argues, is not moral but logical: "Mark that the basic reflection of
morals here is not itself a moral but a logical one: The 'I *can* will' or 'I
cannot will' expresses logical compatibility or incompatibility, not moral
approbation or revulsion."[17] On this interpretation, Kant's imperative is
based on logical coherence.[18] Persons acting against it are contradicting

15. Ibid., 11.

16. Ibid., 10. Cf. Kant, *Groundwork of the Metaphysics of Morals*, 70 (4:402).

17. Jonas, *Imperative of Responsibility*, 11.

18. That Kant's categorical imperative, at least in its first formulation, is entirely
formal no one doubts. However, that for Kant all of morality is simply a question of
logical coherence one does not need to take for granted. Thus for example Bernard
Sève points out how for Kant what is at stake is not merely the logical connection between states of affairs. It is true that the idea that one day humanity cease to exist does
not contain any logical contradiction. And yet, for Kant the question would be about
the *will* to produce such a situation, which according to Sève, would always be impossible on Kant's terms (cf. Sève, "Hans Jonas et l'éthique de la responsabilité," 82–83).

themselves, insofar as, when they are lying, they will to lie but could not want everyone else to lie, as they themselves would like to live in a society where there is truthfulness and where promises exist. The imperative that Jonas proposes needs to be of a different kind, since "there is no self-contradiction in the thought that humanity would once come to an end, therefore also none in the thought that the happiness of present and proximate generations would be bought with the unhappiness or even nonexistence of later ones."[19] A generation that chooses for itself the life of highest excitement and over it either refuses to procreate or in other ways risks the nonbeing of future generations is not logically incoherent.

Rights?

In his search, Jonas also excludes another possible candidate for a basis of his imperative, i.e., rights and duties: "What we require of our principle is not supplied by the traditional idea of rights and duties."[20] Usually, rights and obligations are reciprocal, that is, when we have rights on the part of some, we have obligations on the part of others. Hence, given our assumed obligation for the being of future generations—future generations ought to be, and we ought not to do anything that would impair this being—we might expect a corresponding right to existence on the part of these future generations. As Jonas explains, however, there is a simple and yet insurmountable problem here. To be the actual bearer of any rights whatsoever, one has to exist already. Real existence is a necessary condition for the possession of rights. "The nonexistent makes no demands and can therefore not suffer violation of its rights. It may have rights when it exists, but it does not have them by virtue of the mere possibility that it will one day exist."[21] This consideration does not, as such, preclude modal variations and hypothetical thought experiments, such as, "if A exists, then it has a right to X." But here, existence, even counterfactually, is presupposed. While A can be violated in its right to X so long as it actually exists, its coming into existence cannot be the object of any right. As Jonas puts it, a thing "has no right to exist at all before it in fact exists. The claim to existence begins only with existence."[22] The conceptual pair

19. Jonas, *Imperative of Responsibility*, 11.

20. Ibid., 38.

21. Ibid., 39.

22. Ibid.

rights-duties presupposes a somewhat reciprocal relationship between those who are the bearers of the obligation and those who are the bearers of the rights. The relationship between existing and non-existing things, however, is utterly nonreciprocal.[23] Hence, if we nonetheless say that the existence of future generations obliges our present action—despite their current nonbeing—this obligation must derive from something other than these generations' rights.

An "Ontological" Argument

In his attempt to ground his new imperative, Jonas takes a daring step by resorting to a discipline that Jürgen Habermas, for instance, has declared long since dead,[24] namely metaphysics. For this move, Jonas has received a lot of criticism,[25] and even those scholars who, at least in Germany, can be counted among those most dedicated to his thought do not follow him here.[26] And yet, in his defense we can say that even though in this argument Jonas goes "beyond the physical," he nonetheless remains quite down to earth. Whether or not the argument will be convincing will depend on the degree to which possible interlocutors show themselves open to some primary evidences to which Jonas appeals and which serve him as fundamental premises. This itself is not a weakness of Jonas' line

---

23. Due to the complexity of the problems raised by Jonas' insistence on nonreciprocity, we will reserve a more detailed analysis—and critique—of it to the final section of this chapter.

24. Cf. Habermas, *Postmetaphysical Thinking*.

25. Wolfgang Kuhlmann, for instance, speaks of Jonas' "failed philosophical foundation of his proposed basic norms," which however, does not necessarily invalidate a practical philosophy that is "primarily committed to coping with the concrete crisis" (Kuhlmann, "'Prinzip Verantwortung' versus Diskursethik," 282; translation my own).

Similarly, Johannes Wendnagel calls Jonas' argumentative foundation "untenable" (cf. Wendnagel, *Ethische Neubesinnung*, 7). Referring to Wendnagel's work—and speaking in the defense of Jonas—Nicola Russo remarks that here "it becomes very evident how a reading of *The Imperative of Responsibility*, however subtle, will be very compromised if it is lacking any reference to *The Phenomenon of Life*" (Russo, *La biologia filosofica*, 13n6; translation my own).

26. In this sense, Dietrich Böhler thinks that it is important to ask, "Can the problem of our responsibility for the future, which Hans Jonas has raised, be solved this side of a teleological understanding of nature and a deduction of ethics from ontology?" Answering his own question in the affirmative, Böhler continues then to "seek a direct dialogue only with the moral philosopher and not with the natural philosopher Hans Jonas" (Böhler, "In dubio contra projectum," 244–45; translation my own).

of reasoning, but a predicament shared with every thinkable argument: every argument has to begin with premises that need to be presupposed. Many times, when a partner in the dialogue wishes not to grant given premises, one can argue these premises themselves. For this procedure, however, one will again need premises which must be granted. At some point one will arrive at primary evidences or first principles, which cannot themselves be argued but are the condition for the possibility of arguments as such, as is, for instance, the principle of noncontradiction. In what follows we will present Jonas' argument for the foundation of humanity's obligation to be, beginning, as he himself does, with the conclusion and then working out the premises.

Again, the question is as follows. Jonas, with his new categorical imperative for the technological age, postulates our obligation to ensure the future presence of human beings in the world. "Act so that the effects of your action are compatible with the permanence of genuine human life."[27] This obligation, however, cannot be founded in a Kantian way by reference to logical coherence or performative self-contradiction, nor can it be grounded in the rights of future generations, because there cannot be any right of nonexisting subjects to come into existence. In solution to this difficulty, Jonas himself proposes to ground humanity's duty to be—and with that the present generation's obligation toward future generations—in the idea of humanity itself: "With this imperative we are, strictly speaking, not responsible to the future human individuals but to the *idea* of Man, which is such that it demands the presence of its embodiment in the world."[28] If I understand what humanity means, I will understand that it ought to be, or, in other words, that a world in which humanity is present is better than one in which humanity is not present.

Jonas calls this idea "ontological" because to him it is an "idea of *being*" which claims that its content "*ought to exist.*"[29] He immediately distinguishes it from Anselm's ontological proof, which "concerning the concept of God, guarantees the existence of its subject already with the essence."[30] Most certainly the ontological "idea of Man" does not guarantee the existence of its subject. What it bespeaks is rather the obligation for its subject to be. So, perhaps there is at least some analogy between

27. Jonas, *Imperative of Responsibility*, 11.

28. Ibid., 43.

29. Ibid., 44.

30. Ibid., 43.

the two, after all. Anselm argues that the very idea of God implies the existence of the being of which it is the idea. If I understand what God is, I understand that he exists. The point of the analogy here would then not be the question of existence, but the procedure of arguing from the idea of a thing to a property to be demonstrated. The idea of humanity does not help me to understand that humanity is, but that it ought to be. Anselm's proof, of course, works for someone who knows God's essence, which is only God himself, so that in Aquinas' view, for instance, this proof is not very helpful for us who have a less than perfect knowledge of God.[31] Jonas' "ontological idea" is different here as it only requires us to know something about the nature of humanity and not to have a direct insight into the essence of God. Being humans ourselves, humanity is not as such beyond the grasp of our knowledge. Hence, even though there is some similarity between Anselm's and Jonas' arguments, the latter may still work even if the former does not.

Being, Purpose, and Value

Jonas is quite aware that by grounding his categorical imperative in the idea of humanity, he violates two fundamental contemporary dogmas: (1) that there is no metaphysical truth and (2) that one cannot derive an ought from an is, i.e., that one cannot derive any duty from mere facts. The latter presumption, Jonas argues, presupposes "a concept of being that has been suitably neutralized beforehand (as 'value-free,') so that the nonderivability of an 'ought' from it follows tautologically."[32] If being is inherently neutral, and no other concept of being is possible, then, of course, obligation cannot be derived from it. Now our author makes the case that the assumption *that* being is neutral is itself a metaphysical position, and as such violates the first contemporary "dogma," that is, the prohibition of metaphysics.[33] That there is no metaphysical truth is tautologically true if we presuppose a concept of knowledge that is restricted to natural science. Evidently and necessarily "'scientific truth' is not to

---

31. Thomas Aquinas, *Summa Theologica*, I, 2, 1: "This proposition, 'God exists,' of itself is self-evident, for the predicate is the same as the subject, because God is His own existence as will be hereafter shown. Now because we do not know the essence of God, the proposition is not self-evident to us; but needs to be demonstrated by things that are more known to us, though less known in their nature—namely, by effects."

32. Jonas, *Imperative of Responsibility*, 44.

33. Cf. ibid.

be had about metaphysical objects" if we understand all science to be about physical objects.[34] Jonas concludes: "So long as it is not indisputably shown that this exhausts the whole concept of knowledge, the last word on the possibility of metaphysics has not yet been spoken."[35] He then points out that the ethics he is proposing is not the only one that presupposes metaphysics; in fact, every known ethic does, given the fact that even materialism, for instance, is a type of metaphysics.[36]

In this context Jan Schmidt suggests that the use of the word "metaphysics" for what Jonas is doing here is "imprudent from the point of view of argumentative strategy." There is the danger that people may understand him to mean "a contingent postulate or . . . a comprehensive system of dogmas that is immune to critique and revision," while in fact what he has in mind is "a set of implicit ways of understanding the human person, nature, and science."[37] A metaphysics in this sense will have to prove itself as a "hermeneutics of existence,"[38] and Schmidt is agreed with Jonas that "we will remain in a dialectical relation to any kind of metaphysics," i.e., "the question is not *whether* we have a metaphysics . . . , but *which* metaphysics it can and *should* be."[39]

Having provided his justification for taking recourse to metaphysics, Jonas plunges himself immediately into the biggest metaphysical question: the Leibnizian "Why is there something rather than nothing?"[40] To say that humanity ought to be, that humans, now that they exist, should exist, is to say that it is better for them to be than not to be. This raises the question why it should be better for anything at all to be rather than not to be. Why is it better that something exists rather than that there be nothing at all? And then, is there something about the being of humans that qualifies their existence as particularly "owed"? Jonas points out that there is one thing that by itself would serve to found a claim to being, if we were able to verify its presence, namely value or goodness. "The mere fact of value . . . being *predicable* at all of anything in the world . . . is

34. Ibid.

35. Ibid.

36. Cf. ibid.: "In every other ethic as well . . . a tacit metaphysics is imbedded ('materialism,' e.g., would be one), and they are not a whit better off in this regard."

37. Schmidt, "Die Aktualität der Ethik von Hans Jonas", 549 (translation my own).

38. Ibid., 564n51 (translation my own).

39. Ibid., 549 (translation my own).

40. Cf. Jonas, *Imperative of Responsibility*, 46.

enough to decide the superiority of being . . . over nothingness."[41] Even a world where the scale of value and disvalue tips to the side of disvalue would still be better than no world at all, since only an existing world can have the capacity for value, which "is itself a value, the value of all values, and so is even the capacity for antivalue . . . , insofar as the mere openness to the *difference* of worth and worthlessness would alone secure to being its absolute preferability to nothingness."[42] Even the capacity for disvalue would already assure being's superiority over nothingness because disvalue presupposes that the difference between value and disvalue applies. Thus it is not only this or that value in particular, but already the capacity for value in general that answers to the question of why this or that being, which has this capacity, ought to be.

What Jonas then sets out to do is to ask whether there actually is such a thing as value or goodness in the world, or at the very minimum, whether it is at least conceptually conceivable. In a first step he seeks to clarify the relation between values, goods, and ends. An end is that for the sake of which a thing exists. Thus, the purpose of a hammer consists in hammering. Jonas emphasizes that the ascription of a purpose to a thing does not yet imply a value-judgment; it is simply a question of understanding properly what the thing is. Hence, "I may prefer a natural state without hammers to a state of civilization in which nails are driven into walls,"[43] and yet so long as I describe a hammer correctly, I will acknowledge that this is precisely its purpose. A value-judgment only comes in at a next point: things that have purposes can fulfill these purposes in a better or worse way. There are good and bad hammers, depending on their fitness to the task of hammering. Importantly, however, these value-judgments, too, are nothing subjective or arbitrary but again depend on my understanding, not on my taste; they do not "rest upon value decisions or goal-settings of mine: they are deduced from the perceived being of the respective objects and rest upon my understanding of them, not upon my feelings about them."[44] Thus, our capacity to form the concept of the good is derivative of our ability to see ends in things; "it is the 'good' according to the measure of fitness for an end (whose own goodness is not judged)—thus *relative* value *for* something."[45]

41. Ibid., 48.
42. Ibid., 49.
43. Ibid., 50.
44. Ibid., 51.
45. Ibid.

In a next move, Jonas asks himself about the status of the things that have ends and about the value of these ends. Can we ask whether a certain end is better than another? In other words, is there an "end-in-itself" and is there a "value-in-itself"? As to the question of the end-in-itself, taking up again some of his insights from *The Phenomenon of Life*, Jonas argues that there is at least one kind of being in the world that is its own end: the living organism. Living beings certainly *have* ends; at the very least they try to maintain their lives. To be a living thing means to be a concerned existence, one whose existence is threatened by death and who has a genuine interest in its own survival. This does not make life the highest good or survival the only purpose. All that Jonas is interested in showing is that living beings have at least one purpose: to maintain themselves in being, and in this way they *are* their own ends. With living beings, purpose has made its appearance in the world and with purpose also value. "On the strength of the evidence of life . . . we say therefore that purpose in general is indigenous to nature. And we can say something more: that in bringing forth life, nature evinces at least *one* determinate purpose—life itself."[46] From the perspective of living beings, the world is not value-neutral at all. They are interested beings for which some things are good and some things are bad; they are capable of the distinction between good and bad. Now insofar as the world brought forth living beings, the world itself is purposive and has values.

And yet, to Jonas' mind, we still have to go a step further; we will need the concept of the good: "For real, obligatory affirmation, the concept of the *good* is needed, which is not identical with the concept of value, or if you will, which signifies the distinction between objective and subjective status of value)."[47] While the word "value" ultimately derives from the context of the exchange market,[48] the "good" by its very concept "entails the demand for its being or becoming actual."[49] What, then, is this "good" or "good-in-itself," which deserves to be affirmed on the very basis of its *being*? Where, then, is the bridge between the "is" and the "ought"? For Jonas, it lies in a special capacity that is ontologically rooted and that in itself contains the affirmation of its "ought-to-be," namely

---

46. Ibid., 74.
47. Ibid., 77.
48. Cf. ibid., 83.
49. Ibid., 79.

purposiveness as such: "We can regard the mere *capacity* to *have* any purposes at all as a good-in-itself."[50]

Nevertheless, Jonas raises the following objection: "That the world
has values . . . follows directly from its having purposes"; however, "it is
the prerogative of human freedom to be able to say no to the world."[51]
The human being does not need to agree with the world's "value-judgments" and could still claim that it were better if there were nothing. This
is precisely the problem that Vittorio Hösle points out: "As the factual
values of a society do not bind me just for that, so the factual purposes
of organisms could be rejected by an observer as devoid of any objective
value. . . . Indeed Jonas presupposes that the whole structure of being,
which brings forth entities that strive to convert an ought into an is, has
intrinsic value. Can this presupposition be grounded?"[52] For Hösle, to
argue for the intrinsic value of the world's purposive structure, "Jonas'
theological reflections could become helpful. . . . One possibility would
be to refer to the divine origin of this structure: As created by God, who is
the source of all values, it shares in his value."[53] However, to Hösle's mind
this solution brings with it all the problems inherent to the proofs for the
existence of God and would ultimately be circular: "One needs God to
be able to found the objectivity of ethics; but the main argument for the
existence of God is that one needs him for the sake of the presupposed
objectivity of ethics."[54] In this context, Jonas himself, at least, does not
think that his argument presupposes natural theology. Rather, he makes
the claim that one can separate the idea of the "ought-to-be" of the world
from the idea of its divine authorship. Presumably, God created because
he judged that the world *ought to* be, that it was good for the world to be.
When the biblical creation account tells us that at the end of his works
God saw that it was very good, this implies that "he willed it because he
found that it ought to be." In fact, for Jonas, "the perception of value in

---

50. Ibid., 80.

51. Ibid., 76.

52. Hösle, "Ontologie und Ethik," 120 (translation my own).

53. Ibid. (translation my own).

54. Hösle, "Ontologie und Ethik," 120–21 (translation my own). In his view, a more
promising way of founding ethics—presumably one that does avoid questions of metaphysics—is the transcendental-pragmatic one along the lines of Karl-Otto Apel and
Jürgen Habermas: "One tries to show that certain moral principles are already implicit
in the argumentative situation" (Hösle, "Ontologie und Ethik," 120–21; translation my
own).

the world is one of the motives for inferring a divine originator (formerly even one of the 'proofs' for God's existence), rather than, conversely, the presupposition of the originator being the reason for according value to his creation."[55]

In his own argument for purposiveness as good-in-itself, Jonas distinguishes between two possible stances one could take in front of it: its negation and its affirmation, which raise different questions. *Denying* the goodness of purposiveness is impossible as long as one holds a monistic understanding of the world, while *affirming* it ultimately requires a direct insight. As to the first, in the very negation of the world as it is—namely purposive—a person inevitably needs to make reference to some standard, insofar as he or she implies that it would be *better* if things were different or if there were no things at all. This standard then would have to be outside the world—since the world itself is denied—and we would find ourselves again in some form of dualism: "I can *legitimately* dissent from nature only if I can appeal to a tribunal outside it—that is, to a transcendence which I believe to possess the authority I deny to the former: hence, only by embracing some kind of dualism. (A dissent based upon mere taste or mood would be frivolous.)"[56] As long as we would like to maintain a monistic ontology—for which, as we have seen, there are good reasons—a wholesale dissent from the "value-judgments" of nature will not be possible.

However, that purposiveness in itself is not only a *value* that cannot be meaningfully denied but that it is a *good* that at the same time must

---

55. Jonas, *Imperative of Responsibility*, 48. Even if we can grant Jonas this point, the question of whether God exists is not at all indifferent for our conduct. As Pope Benedict XVI points out in his address at the Hofburg in Vienna, "The issue is whether reality originates by chance and necessity, and thus whether reason is merely a chance by-product of the irrational and, in an ocean of irrationality, it too, in the end, is meaningless, or whether instead the underlying conviction of Christian faith remains true: *In principio erat Verbum*—in the beginning was the Word" (Benedict XVI, *Address during the Meeting with the Authorities and the Diplomatic Corps*, Hofburg, Vienna, September 7, 2007). But if the world were simply the product of the irrational, then moral action would be impossible. Life would be without reason or rhyme and no criteria for good or bad could be given. This is why, still as a cardinal, Joseph Ratzinger proposed the following consideration: "The attempt, taken to the extreme, to mold human things by doing completely without God brings us closer and closer to the edge of the abyss, towards the total setting aside of man. So we should turn the Enlightenment axiom on its head and say: even those who cannot find the way to acceptance of God should try in any case to live and direct their lives *veluti si Deus daretur*, as though God existed" (Ratzinger, "Europe in the Crisis of Cultures," 41.

56. Jonas, *Imperative of Responsibility*, 77.

be affirmed in an obligatory way, i.e., that it is a good that makes a claim on my will to respect it and to further it, this for Jonas is the result of a primary intuition. Thus, "we grasp with intuitive certainty that it [purposiveness] is infinitely superior to any purposelessness of being."[57] For Jonas, the superiority of purposiveness over its absence cannot be argued any further but is immediately evident.

As we have asserted above, in the last analysis, appeal to primary evidences cannot be avoided. Every argument needs to begin with some presuppositions that need to be granted, all the way down to presuppositions that cannot themselves be argued anymore, insofar as they are the conditions for the possibility of arguments—such as the principle of non-contradiction, which one cannot argue against without presupposing it at the same time. The question is whether Jonas' principle of purposiveness as intuitive good-in-itself indeed enjoys such a high evidence. Wolfgang Kuhlmann, for instance, remains unconvinced. He writes, "Finally [Jonas] claims that evidently—with an evidence that cannot be argued any further—the capacity to have purposes must be regarded as a good-in-itself, a good from which formally follows the ought, the obligation. . . . It is entirely unclear, how, using this—as he himself calls it—metaphysical-ontological value theory, Jonas can claim to have bridged the gap between being and obligation."[58] As we have mentioned above, Vittorio Hösle, too, is not entirely persuaded by Jonas' metaphysical reflections, and yet in the context of the present discussion he wants to come to Jonas' aid by arguing that indeed the affirmation of the existence and value of purposes forms part of the transcendental conditions of arguments. "In particular it seems to me that Jonas' argument that the structure of purposes as such is axiologically superior to the structure of nonpurposes can be grounded by the transcendental-pragmatic reflection that arguments themselves have a teleological structure that is always already presupposed when we try to contest them. Thus the argument that contests purposes has itself a purpose—namely to show that purposes either do not exist or in any case do not possess any special value."[59] In other words, by engaging in an argument, persons who seek to dispute the existence of purposes or of

---

57. Jonas, *Imperative of Responsibility*, 80.

58. Kuhlmann, "'Prinzip Verantwortung' versus Diskursethik," 282 (translation my own).

59. Hösle, "Ontologie und Ethik," 121 (translation my own). Cf. also McKenny, *To Relieve the Human Condition*, 59: "A denial of purposiveness as good . . . itself expresses purpose."

their value imply that they have a purpose themselves—namely to defeat the opposing position—and that they think it *worth* their while to pursue this purpose.

Responsibility

Let us grant Jonas' argument and acknowledge that at this very foundational level of reflection we may not be able to avoid recourse to some primary evidences, which, however, may be more evident to some than to others.[60] So far this argument only shows, however, that a world in which there are purposes, or, put differently, in which there are living beings, is better than one in which there are none, and that hence this world has a claim to being, that it *ought to* be. What, however, is the special place of human beings in the world? How can we say that humanity ought to be in particular and not just that living beings ought to be? For Richard Bernstein, this is a crucial point in the argument where Jonas fails to provide a proper rationale for what he is doing: "It is never entirely clear what warrants the move from the imperative to preserve organic life to the more specific conclusion that the *primary* object is to preserve human existence."[61] For Jonas, what distinguishes human beings and what

---

60. Gilbert Hottois, for example, challenges the very notion that being is superior to nonbeing, criticizing Jonas' argument because to his mind it "necessarily disvalues" the latter. He calls on the testimony of "all the philosophers, from dialectical Idealism to Existentialism or to critical rationalism, who would ascribe to nonbeing, to negation, a principal role, essential as the font of freedom, of becoming, of progress, of meaning itself. . . . One cannot see how being alone can make possible the distinction between good and bad, between the more and the less, between value and nonvalue. These distinctions presuppose negation and hence nonbeing. In other words, the notion of value equally needs being *and* nonbeing to work" (Hottois, "Un'analisi critica del neo-finalismo," 101–2; translation my own).

That negation is necessary for distinction is undeniable and indeed no new thought. Thus for the medievals it is the very mark of creatureliness to fall short of being in its fullness. While for God essence and existence fall into one, a creature's essence is defined by its contingency, its very lack of being. This, however, does not make nonbeing in itself a value. If it has any "value," then only in reference to being—the being that it lacks which makes for an entity's distinction. Ultimately, it seems that Hottois has not grasped the radicality of Jonas'—or Leibniz'—question. It is not about being and the negation of being still rooted in being, but rather why there should be anything at all and not nothing. What is at stake here is *being as such* and why *it* is better than nonbeing "as such," which probably however cannot even be thought.

61. Bernstein, "Rethinking Responsibility," 18.

makes them "the supreme outcome of nature's purposive labor"[62] is the fact that they alone have the capacity to respond to the claim that issues forth from being. "There would be no 'thou shalt' if there were no one to hear it and on his own attuned to its message."[63] Human beings have moral sentiment with which they can perceive the demands of being, and they have a free will with which they can say yes or no to these demands. This qualifies them to be bearers of *responsibility*. In this context Gerald McKenny summarizes Jonas' position in the following way, "Only human beings can have responsibility for the purposes of other beings. This capacity means that they can in a unique manner include fellow human beings in their responsibility. To be responsible for fellow humans is to be responsible for those who are themselves responsible."[64] It is this concept of responsibility which for Jonas is the distinguishing mark of humanity and the foundation of its duty to be. In what follows, we will look at this idea in more detail.

Jonas distinguishes between responsibility in a legal or moral sense, that is, as the imputability of actions or their consequences to their author and responsibility as an original moral sentiment—the "feeling of *responsibility*"—that is directed to what is both fragile and under one's power and hence entrusted to one's custodianship[65]: "There can be 'responsibility' . . . in two widely differing senses: (*a*) responsibility as being accountable 'for' one's deeds . . . ; and (*b*) responsibility 'for' particular objects that commits an agent to particular deeds concerning them."[66] Thus, responsibility in the first sense means that I am legally or morally responsible for my actions and their consequences because I am their author. With legal responsibility, the emphasis is on responsibility for

---

62. Jonas, *Imperative of Responsibility*, 82.

63. Ibid., 86.

64. McKenny, *To Relieve the Human Condition*, 61.

65. Cf. Jonas, *Imperative of Responsibility*, 87: "The object of *responsibility* is emphatically the perishable *qua* perishable. . . . Yet just this far from 'perfect' object, entirely contingent in its facticity, perceived precisely in its perishability, indigence, and insecurity, must have the power to move me through its sheer existence. . . . And it evidently has this power, or else there would be no *feeling* of responsibility for such an existence." For responsibility as a moral sentiment, see also Jonas, *Imperative of Responsibility*, 92: "The dependent in its immanent right becomes commanding, the power in its transitive causality becomes committed, and committed in the double sense of being objectively responsible for what is thus entrusted to it, *and* affectively engaged through the feeling that sides with it, namely 'feeling responsible.'"

66. Jonas, *Imperative of Responsibility*, 90.

the consequences. Even if I cause injury quite involuntarily, I have to pay restitution; "it suffices that I was the active cause."[67] With moral responsibility, the emphasis is on the action itself. If I committed a morally blameworthy deed, not only do I need to pay restitution for any possible negative consequences but deserve punishment for the act itself. As Jonas puts it: "It is more the deed than the consequence which is punished in the case of a crime, and it is according to the deed that the measure of expiation is fixed."[68]

For Jonas the distinction between legal and moral responsibility "is reflected in the distinction between civil and criminal law."[69] Yet both have in common that they refer to the imputability of the action to its author. In contrast, what Jonas is much more interested in is responsibility in the second sense, which he calls a moral sentiment, and which is the attitude we owe to those entrusted to our care. "The well-being, the interest, the fate of others has, by circumstance or agreement, come under my care, which means that my control *over* it involves at the same time my obligation *for* it."[70] Thus, there are two conditions for this kind of responsibility, vulnerability and power: the vulnerability of the one entrusted to it and the power of the one who has it. For this reason for Jonas the relation of responsibility is inherently a nonreciprocal one. As a subject of responsibility I have power over the object of responsibility, which "lies outside me, but in the effective range of my power, in need of it or threatened by it."[71] *Objects* of responsibility can only be beings that are threatened in their existence, beings that have something to lose because they are interested in something in the first place. In other words, only living beings can be the objects of responsibility in the proper sense.[72] Here the conclusion of the whole previous argument about value and purpose in nature enters as a premise. Responsibility is possible only because there are possible objects of it, beings whose mode of existence is concern and who are dependent on other beings for their continued existence.[73] Pos-

67. Ibid., 90.

68. Ibid., 91.

69. Ibid.

70. Ibid., 93.

71. Ibid., 92.

72. Cf. ibid., 98: "Only what is alive, in its constitutive indigence and fragility, *can* be an object of responsibility."

73. Jonas raises the question of a possible responsibility of an artist for his artwork. He acknowledges that "there is such a thing as that hard-to-define, in its own way

sible *subjects* of responsibility can only be beings that have power over others to further and promote or to obstruct and spoil their concerns and interests. Only human beings are such subjects of responsibility because only they have such power: "Every living thing is its own end . . . . In this, man has nothing over other living beings—except that *he* alone *can* have responsibility *also* for them, that is, for guarding their self-purpose."[74] This power, in turn, is rooted in their freedom. While a lion certainly has "power" over the gazelle, it does not have responsibility for it because the lion is not free to hurt or not to hurt the gazelle. Lacking freedom, the lion only sees in the gazelle a possible object of food, not of responsibility. Since human beings, in contrast, have the power to hurt or promote, to do or not to do, they are free and hence *can* be responsible, which for Jonas immediately implies that they *must* be responsible: "An 'ought' is concretely given with the very existence of man; the mere property of being a causative subject involves of itself *objective* obligation in the form of external responsibility."[75]

Though in principle responsibility can be directed to everything that is alive, for Jonas it is primarily something among human beings. Here, every actual relationship of responsibility is nonreciprocal; however, "on principle it is reversible and includes possible reciprocity."[76] In theory, at least, tomorrow I could become responsible for a person who is responsible for me today. Besides, by the very fact of human lack of self-sufficiency, one can say with Jonas that on a general level responsibility is reciprocal after all, but only in the sense that everyone is both at the same time responsible for others and the responsibility of others, although these "others" are different.

To Jonas' mind the paradigmatic cases of responsibility are the responsibility of parents for their children and the responsibility of the head of state for his or her nation. While there are many important differences between these two, what they have in common is that both the parents and the head of state are obliged to the welfare of those entrusted

---

highest, 'responsibility' of the artist for his work." However, while the artwork is a testimony to human transcendence, insofar as it has become part of the human world, which it is usually intended to become, it "exists after all for men only, for their sake, and only as long as such are around. The greatest masterpiece becomes a mute piece of matter in a world without men" (ibid., 100–101).

74. Ibid., 98.
75. Ibid., 99.
76. Ibid., 98.

to their care. In the case of the parents, this obligation was enjoined on them without any prior choice on their part; it is a natural obligation based on the mere existence of the child. Responsibility in this case is unconditional and cannot be abnegated. In the case of the head of state, in contrast, the obligation is contractual and grounded in a prior agreement which leaves open possibilities of being released from it.[77] In both cases the responsibility is nonreciprocal. But particularly the parent-child relationship is the paradigmatic image of responsibility as an obligation to custodianship that issues from the very being of the other. In fact, the newborn "is an *ontic* paradigm in which the plain factual 'is' coincides with an 'ought' . . . . Here the plain being of a *de facto* existent immanently and evidently contains an ought for others," namely the obligation to care for him or her.[78] For Jonas, then, responsibility is ultimately rooted in the call of being itself, which issues forth from singular, vulnerable, and at the same time precious beings.

Having discussed what responsibility is, what its conditions are and having demonstrated with the help of examples that it really does exist, Jonas takes a final step in his argument for the "ought-to-be" of humanity. Why, in other words, is the human being the primary object of human responsibility? Why do we need to be responsible for human beings? Jonas denies that this has something to do with their "greatness." If we were to compose a balance sheet with human achievement and virtue on the one side and human atrocities on the other, for Jonas the scale would most likely tip toward the latter: "The price of the human enterprise is, in any case, enormous; man's wretchedness has at least the measure of his greatness; and on the whole, I believe, the defender of mankind, in spite of the great atoners like Saint Francis on his side, has the harder case."[79] Our duty, thus, is not toward actual human achievement. Rather, we are ultimately responsible for human beings because they are moral beings, in the sense that they are capable of being moral or immoral, that they are capable of being responsible or being irresponsible. Now for Jonas, "the possibility of there being responsibility in the world, which is bound to the existence of men, is of all objects of responsibility the first."[80] The duty for humankind to exist derives from the fact that responsibility,

77. Cf. ibid., 95.
78. Ibid., 130–31.
79. Ibid., 99.
80. Ibid.

once it has come to exist, ought to continue to be. Given that the idea of human persons includes the possibility of responsibility, they ought to be: "The naked ontic fact of their existing at all, in which they had no say, becomes for them the ontological command: that they continue to be such."[81] Thus, the title of Jonas' work—at least in English—takes on a perhaps unexpected meaning. Speaking of "the imperative of responsibility," our author does not only and simply mean that we ought to be responsible, which no one would deny, but that *responsibility ought to be*. Our duty is toward the future possibility of responsibility which is part of the idea of humanity.[82] Humanity ought to be because with it the principle of responsibility has entered the world, and a world in which there is responsibility is better than one in which there is none.

That for Jonas our responsibility toward the future of humanity is primarily grounded in our responsibility toward responsibility itself becomes also very clear in a passage where he engages in a thought experiment. Suppose we were in some way to manipulate our descendants in a manner that violates their dignity but that at the same time makes them feel quite happy, so that blaming us for their lot would not even occur to them. If they agreed to their condition, which we have set in place, how could it be said that we have violated any of their rights? Here Jonas warns us that "such an agreement might be the last thing we should wish for a future humanity, namely, if it were purchased with the dignity and vocation of man."[83] People's dignity, however, derives from their being capable of responsibility, from their being moral beings. Hence, the first question we have to ask is not about our descendants' *rights;* rather, it is "their *duty* over which we have to watch, namely their duty to be truly human: thus over their *capacity* for this duty—the capacity to even attribute it to themselves at all—which *we* could possibly rob them of with the alchemy of our 'utopian' technology."[84] In the imagined scenario, the very lack of blame—our descendants' incapacity to identify their condi-

---

81. Ibid., 99–100.

82. Here, Gerald McKenny, commenting on Jonas, points out that while responsibility cannot mean that other people's "individual purposes become ours, since these will differ from ours, . . . what we can uniquely include in our own responsibility is the responsibility of others: their very capacity for exercising responsibility according to the demands of situations that call for it" (McKenny, *To Relieve the Human Condition*, 61).

83. Jonas, *Imperative of Responsibility*, 41.

84. Ibid., 42.

tion as deprived—would be to our blame. "It could be that we would rather have to accuse ourselves of the fact that *no* accusation against us issues hence."[85] Thus, our first responsibility is to the continued existence of responsible beings, of human beings as moral beings. Jonas' new categorical imperative, "Act so that the effects of your action are compatible with the permanence of genuine human life," is hence based on the idea of human beings as beings capable of responsibility.[86] That the capacity for responsibility, once it has appeared, ought to continue to be, is something that according to Jonas can immediately be seen the moment we understand what responsibility is.

To Richard Bernstein's mind, at least, this last step is actually far from obvious. Agreeing with Jonas' conclusions, he finds fault with his way of arriving at them. Here he makes the good point that one can meaningfully ask, "Why is the primary object of responsibility to preserve the possibility of being responsible in the world?"[87] Bernstein is concerned about "a crucial gap" in Jonas' argument: "For if one is to move from the premise that there is a supreme obligation to preserve the conditions for the possibility of organic life to the conclusion that the 'existence of mankind comes first,' then one needs an independent argument to justify this strong conclusion. It is this argument that I find lacking in Jonas—the argument (not the mere assertion)"[88] Hence, while to Bernstein's mind Jonas was able to make a good case for our duty to preserve organic life, the argument for the human being's special place is still wanting, although it is strongly needed, because, "ironically, on Jonas' own premises, if there were no human beings there would be no serious threat to the continuation of life on this planet."[89] It seems that Bernstein is right when he suggests that the reason Jonas does not draw any hideous conclusions from this fact is simply that he "is tacitly appropriating a great deal from 'traditional ethics.'"[90] If our responsibility were simply about "preserving

---

85. Ibid., 41.

86. Dietrich Böhler clarifies that when Jonas speaks of our responsibility for "the idea of man," what he has in mind "is more than the pure fact of human existence, namely . . . 'genuine human life,' which as such implies capacity for morality and capacity for responsibility" (Böhler, "Zukunftsverantwortung, Moralprinzip und kommunikative Diskurse," 244; translation my own).

87. Bernstein, "Rethinking Responsibility," 18.

88. Ibid.

89. Ibid.

90. Ibid., 19.

the continuation of life—even human life—on this planet," then one can think of many ways of doing so "which we would find morally repulsive." From this Bernstein concludes that "Jonas's 'ethics of responsibility' demands the acknowledgment of other supplemental imperatives and principles that can orient our choices and actions."[91]

Another objection one can raise against Jonas' procedure involving a long metaphysical argument, is that—independent of what one thinks of metaphysics—it is simply unnecessary. Thus Wolfgang Kuhlmann claims that "Jonas' main merit lies in his apt articulation and in his convincing recourse to shared pre-theoretical moral intuitions. This is evident especially in the casuistic individual studies of the volume supplementary to *The Imperative of Responsibility,* namely his work *Technik, Medizin und Ethik,* and it is confirmed by the kind of resonance which Jonas' attempt has abundantly received, in which the reaction to Jonas' foundational program hardly plays any role."[92] It may indeed seem that in some ways his conclusions stand on firmer ground than his premises. Wondering why this could be so, we do not intend to take any exception here at his resorting to metaphysics. What to our mind however could have helped Jonas to make his argument appear less forced would have been a clearer distinction between the workings of practical and speculative reason.[93] Of course, human reason is just one faculty, which however works differently depending on whether it attempts to discern what is the case or what is to be done. Speculative reason in its search for what is the case begins with facts and ends with facts. Practical reason, in contrast, seeking what is to be done, has as its premise the good and as its conclusion an action.[94] Now it is true that in some ways the presence of the good in the world is itself a fact, and hence one can meaningfully say, for instance, "It is a fact that life is a good." Jonas constructs his argument for the "oughtness" of humanity the way one does a speculative syllogism, arguing from the fact of the existence of goodness or responsibility to the imperative of its existence. He has an is-statement in his premises and an ought-statement in his conclusion. However, when I speak of the good

91. Ibid.

92. Kuhlmann, "'Prinzip Verantwortung' versus Diskursethik," 300 (translation my own).

93. Cf. for instance, Melina, "La prudenza secondo Tommaso d'Aquino," and Melina, *La conoscenza della morale.*

94. Cf. Aristotle, *On the Soul,* III, 11; *On the Movement of Animals,* 7, and *Nicomachean Ethics,* VII, 3.

as of a fact, for instance, when I say, "It is a fact that life is the object of certain beings' desire," I have not yet seen life in its goodness. It seems that Hume is right and that the naturalistic fallacy is indeed such—from statements of sheer fact there is no bridge to statements of obligation.[95] But as originally as we know about certain facts, we know about certain goods—not in virtue of some perfuse "intuition" but simply in virtue of what we are: teleologically structured beings who have certain interests in virtue of what they are. The good is never given to us in its mere factuality as if it were simply a thought. The good is never just a thought but rather also always the object of desire.[96] Thus, as Jonas himself keeps emphasizing, for a living being, life is affirmed existence. Life is never just a fact but a good to be pursued and to be preserved. The same holds true for other, more specifically human goods, such as sexuality, sociality or truth.

Insofar as humans are rational beings they can go beyond their own teleology and acknowledge and affirm the teleology of other beings. In this way we are able to be affected by the primary evidence of a newborn child, which Jonas adduces. In this child we indeed perceive a case in which the "is" and the "ought" fall together because we know that life is a good for it and that it cannot achieve this good without our help. We can pursue its good as if it were our own, which is possible for rational beings capable of transcending themselves. For Spaemann, this means that for us "the reality of the other in its teleology" can become a motive for action, insofar as we are capable of delighting in other persons' happiness, as Leibniz put it.[97] But we may also think of William James, who claims that the attitude of coming-to-the-aid-of is really the natural attitude of a rational being, that is, the burden of proof, in any case, lies with not-coming-to-the-aid: "Take any demand, however slight, which any creature, however weak, may make. Ought it not, for its own sole sake, to be

---

95. For a helpful response to Hume's challenge, see Melina et al., *Camminare nella luce dell'amore.* 118–23 and also: Pérez-Soba, "*Operari sequitur esse?,*" 65–83.

96. Cf. Melina, "La Prudenza secondo Tommaso d'Aquino," 386: "How can thought generate action? The answer is simple: the conclusion of a syllogism can be an action—and not just a thought on action—only if the very premises of the syllogism are not only a thought but are from the start interwoven with appetite, which moves them. The practical syllogism is thus a syllogism in which from the beginning (major premise), and then always down to the end of the process (conclusion), mind and appetite are interwoven and together work as a single principle that spurs to action" (translation my own).

97. Spaemann, *Happiness and Benevolence,* 78; cf. Leibniz, *Codex Iuris Gentium (Praefatio) (1693),* 171: "Love . . . signifies rejoicing in the happiness of another."

satisfied? If not, prove why not."[98] We may wonder then, whether Jonas' ontological argument for the "ought-to-be" of humanity is ultimately as superfluous as the ontological argument for the existence of God: only those who already know the conclusion will find the argument convincing. In the latter case, that is only God himself, in the former case, this is everyone who is open to perceiving the call of the good.

Some Practical Principles of the Ethics of Responsibility

Jonas' "metaphysical" argument for the "ought-to-be" of humanity forms a central part of his book, but its contribution is not exhausted by this, so that his work can make for fruitful reading even on the part of those who still remain convinced of the "death of metaphysics." It was not Jonas' metaphysical speculations that made his book successful, but his very concrete practical wisdom when it comes to the question of what we should do if we grant that for whatever reason humanity is something to be preserved. How can we act responsibly, how can we take responsibility for those entrusted to our care and in particular for future generations? Here, we will now turn to some of the concrete practical advice Jonas gives for how we should deal with the challenges posed to us by modern technology, through the advance of which the very permanence of future humanity has for the first time become the object of our responsibility, since with this technology for the first time we actually have a power that can threaten human survival. A more theoretical critique of utopianism will also be part of that discussion, since, for Jonas, modern technology is characterized by some inherent utopian tendencies.[99]

The Recognition of Ignorance

What should we do then? As we have already mentioned, the first thing is to admit the limitation of our ability to predict the outcome and future consequences of our new kinds of intervention into nature. Our *power* to effect changes by far outstrips our *knowledge* of the mediate let alone

---

98. James, "Moral Philosopher and the Moral Life," 249.

99. Cf. Jonas, *Das Prinzip Verantwortung*, 9: "In its tendency, though not in its program, the world-wide dynamics of technological progress harbors as such an implicit utopianism" (my own translation from the Foreword to the German edition. The phrase did not make it into the Foreword to the English one).

remote consequences. Hence, the admission of our ignorance is among the first imperatives of Jonas' new ethics: "The gap between the ability to foretell and the power to act creates a novel moral problem. With the latter so superior to the former, recognition of ignorance becomes the obverse of the duty to know and thus part of the ethics that must govern the evermore necessary self-policing of our outsized might."[100]

The Exclusion of the Va-banque *game*

Given our ignorance of its consequences, our action will always be risky. The risk has always been an attribute of every kind of action.[101] What is different, however, with the advance of technological action is that the stakes have become much higher, so that we might even endanger humanity as a whole. Hence in every concrete decision, we must be quite conscious of what is at stake and what we are betting for. In some ways the general rules of every game of chance apply. We can only bet what we can afford to lose, and we must weigh the stakes and the promises. While with small stakes we can allow for many misses, "we may allow but few where greater things are concerned. And in the really great, irreversible ones, which go to the roots of the whole human enterprise, we really must allow none."[102] In other words, we must never play the *va-banque* game, no matter how great the promises, because we must not lose humanity. With human nature, which is capable of "truth, valuation, and freedom," we have something infinite to "preserve in the flux, but something infinite also to lose."[103] On the ground of these considerations, Jonas' practical advice is the following: "The prophecy of doom is to be given greater heed than the prophecy of bliss."[104] This imperative may seem to convey

---

100. Jonas, *Imperative of Responsibility*, 8.

101. Cf. Blondel, *Action*, 4: "We do not go forward, we do not learn, we do not enrich ourselves, except by closing off for ourselves all roads but one and by impoverishing ourselves of all that we might have known or gained otherwise. . . . I must commit myself under pain of losing everything; . . . Practice, which tolerates no delay, never entails perfect clarity; the complete analysis of it is not possible for a finite mind. . . . I cannot put off acting until all the evidence has appeared, and all evidence that shines before the mind is partial. Pure knowledge is never enough to move us because it does not take hold of us in our entirety. In every act, there is an act of faith."

102. Jonas, *Imperative of Responsibility*, 31.

103. Ibid., 33.

104. Ibid., 31.

sheer pessimism. However, as we have said, there is not only something to gain but also something to lose. For Jonas, the person who so little treasures the heritage (including the biological one) that has been passed on to us that he is willing to run every risk to improve it, including the risk of losing everything, is the greater pessimist.[105]

It is true that, as Bertholt Brecht says, someone staying in a burning house should not ask whether it is raining outside.[106] And indeed, speaking of a head of state confronted with a decision about war and peace, Jonas admits the possibility of a situation when one has to risk it all: namely when everything, without one's prior choice, has been put at stake already—that is, in the face of final doom.[107] Here, running the great risk can be licit, but "only under the threat of a terrible future," i.e., in order "to prevent a supreme evil," an evil that we cannot possibly live with. On the other hand, it will always be possible to "live without the supreme good."[108] A few pages later, however, Jonas speaks of the prohibition "to incur the risk of nothingness, that is, to allow the presence of its possibility among the chances of our choice. It [our ethical principle] forbids, in short, any *va-banque* game in the affairs of humanity."[109] For him, there is a difference whether it is a single nation that is at stake or humanity as a whole. In the latter case, a *va-banque* game must always be excluded.

And yet, to our mind, whether for Jonas it could *ever* be licit to risk the being of humanity as such is a question that can still be debated, and perhaps Jonas is not entirely clear here. In some ways, what he says about the head of state and the nation would also apply to humanity as a whole when there is some external and definitive threat of annihilation— for instance by a meteorite. To counter this threat, would it be licit to use means such as extremely powerful nuclear weapons that themselves carry the risk of destroying humankind? When Jonas says, "Never must the existence or the essence of man as a whole be made a stake in the hazards of action,"[110] he invites the question of what should happen in case the *essence* of humanity can only be preserved by risking its *existence*. If Jonas is too strict here with forbidding the wager, he may in fact be

---

105. Cf. ibid., 34.

106. Cf. Brecht, "Buddha's Parable of the Burning House," 290–92.

107. Cf. Jonas, *Imperative of Responsibility*, 36.

108. Ibid.

109. Ibid., 38.

110. Ibid., 37.

inconsistent with his own thought, since for him it is the essence of humanity—namely, the capacity for morality—that is the basis of its duty to exist. If the very reason for why humanity ought to be were at stake—and supposing that it could ever be definitively lost—it seems that humanity itself could be risked, since after that reason has been lost, there would be no more duty for it to be anyway. Thus, it would seem better to risk annihilation than to destroy the capacity that makes human life worth living. In the evaluative section of this chapter we will still deal with this issue in more detail, showing how it does indeed pose a significant problem for Jonas' thought.

Where Jonas leaves no doubt, however, is that it can never be licit to risk everything simply in order to improve one's lot, that is, for the sake of a great future promise. While one cannot live with the supreme evil, "one can live without the supreme good."[111] Our author proposes the following considerations to those who would seek to enhance the human condition even at the risk of losing everything. What is the enhancers' qualification to know what counts as a genuine improvement and how do they know how to achieve it? If they are qualified to know, then the heritage of humanity, from which they themselves come, cannot be all bad, and hence it is worth preserving as it is, even if one can imagine still better states, which, however, cannot be achieved without running severe risks. If, however, this heritage is so bad that it is not worth preserving and that it is worth running every risk at the prospect of improving it, then the enhancers themselves, insofar as they derive from such faulty stock, cannot be qualified to know what is truly good, what is better, and what are the ways to achieving it.[112] If the house were truly burning, then these people would in any case not be the ones to lead us out into the rain.

But Jonas can still imagine a third alternative: Scientists may say that their research and technology are not at all geared toward improving the human condition and that they have no opinion in either way on its goodness or faultiness. They may say that all they mean to do is to play. To Jonas, this latter attitude is the "standpoint of nihilistic freedom." And while this position is not in itself logically incoherent, he sees no need to discuss it, since "we will certainly not entrust our fate to professed irresponsibility."[113]

111. Ibid., 36.
112. Cf. ibid., 32–33.
113. Ibid., 33.

A Heuristics of Fear

Given that we must not lay the heritage of our humanity on the line—that the *va-banque* game must be excluded even if this means renouncing potentially very desirable prospects and promises—how can we know that humanity is actually at stake here or there; how can we know when the stakes are too high? Here Jonas proposes what he calls a "heuristics of fear."[114] By this he does not mean that we should let ourselves be guided by dread or apprehension. What he wants to propose is that we actively exercise our imagination to see what our actions could lead to and that we train our sentiment to let ourselves be affected by what we so imagine. Jonas refers to Aldous Huxley's negative utopia *Brave New World* as an example for what he has in mind here.[115]

It appears that the interpretative function of Jonas' heuristics of fear is twofold. Firstly, it helps us to see our present actions in the light of their possible future consequences. The matter is not about predicting with certainty that such and such will happen. Rather, "it is the content, not the certainty, of the 'then' thus offered to the imagination as possible, which can bring to light, for the first time, principles of morality heretofore unknown for lack of actual occasions to which they could apply."[116] In this case, the prophet does not aim at predicting the future, but rather at making sure the future will not be as predicted.

Secondly, the heuristics of fear also helps us to understand better the object that is threatened. Jonas argues that somehow we are made in a way that we can much more easily recognize what we do *not* want than what we actually *do* want. Imagining the threats to ourselves and our humanity, we come to realize how we would not want to be, and hence we also learn something positive about how we do want to be.[117] Mainly finding out what we do not want may seem little, and Jonas admits to this when he says, "the heuristics of fear is surely not the last word in the search for goodness." Nonetheless, to his mind "it is at least an extremely useful first word."[118] Bernard Sève in effect goes so far as to call the "heuristics of fear" one of the most original ideas of the whole book,[119] and the fact that

114. Cf. ibid., 26–30.

115. Cf. ibid., 30.

116. Ibid., 29.

117. Cf. ibid., 27.

118. Ibid.

119. Cf. Sève, "Hans Jonas et l'éthique de la responsabilité," 76.

Jonas returns to it toward the end of his work speaks in favor of its importance.[120] For Giuliano Sansonetti it is "in any case the idea that most contradicts the common practice, supported by the constitutive principle of the modern epoch, namely the principle of unlimited progress, of the 'magnificent lot and advancement' of humanity."[121] It is precisely this idea of progress that Jonas will take on in his critique of utopianism to which we shall now turn.

Jonas' Critique of Utopianism

Jonas presented these practical considerations at the beginning of his book, then seeking their foundation in his new categorical imperative, which in turn is grounded in an ethics of responsibility and has its ultimate foundation in the claim of being on our wills. In what to our mind can really be considered the second part of his book—beginning with chapter 5 entitled "Responsibility Today: Endangered Future and the Idea of Progress"—Jonas turns to political questions. Given that humanity is endangered, given that it is to be preserved, and given that we know why and even—to some extent—how it is to be preserved, which political system is best suited to guarantee the future of humanity? Is it capitalism or Marxism, or still something else? In his analysis he finds that indeed both capitalism and Marxism rely on the idea of technological progress. In its own theoretical account, Marxism has even made technological progress part of its utopian ideal, which is the reason why Jonas dedicates most of his attention to it. Now, we may be tempted to dismiss his analysis, proposed in the late 1970s, as out of date and irrelevant for us today, given that most communist regimes in the world have since collapsed and those still existing are either on the verge of collapse (North Korea and Cuba) or have basically given up some of their most fundamental communist ideals (China).

To this objection we can respond in two ways. First, it is far from obvious that Marxism, at least as an idea, is dead. People may claim that the "real existing communism" in the so-called communist countries had so little to do with Marx's ideas that the demise of these communist regimes does very little to prove his ideas wrong. Indeed, there continue to be

---

120. Cf. Jonas, *Imperative of Responsibility*, 202.

121. Sansonetti, "Un'etica della responsabilità: Hans Jonas," 481 (translation my own).

communist parties in a number of European countries, such as Germany and Italy, and in some they are growing in influence.

Secondly, Jonas himself admits that his discussion of Marxism would have been too extensive and detailed if it were not for its close alliance with technology and at the same time for the utopian tendencies inherent to our own day's idea of technological progress. As Jonas puts it, "Marxist utopia . . . served as an 'eschatologically' radicalized version of what the worldwide technological impetus of our civilization is moving toward anyway. Thus the critique of utopia is implicitly a critique of technology in the anticipation of its extreme possibilities."[122] In other words, Jonas' critique of Marxist utopianism can still speak to us today, independent of the present or future state of Marxism, because it is at the same time a critique of the utopianism that is built into the very ideal of technological progress, insofar as this ideal implies, as any utopia does, that humanity as we know it is not yet the "true" humanity and that the authentic human being is yet to be realized. McKenny succinctly formulates it this way, "Jonas seeks to identify an arc of thought stretching from Bacon to Marx to contemporary technology, . . . which is summarized in the utopian claim that humanity is yet to be realized in its fullness, a fullness to be brought about in large part by technology."[123] In his section on utopianism, Jonas sets out to examine the implications of this view. Here his main interlocutor is the Marxist Ernst Bloch who has written the book *Das Prinzip Hoffnung* (*The Principle of Hope*).[124] It is precisely in response to this *Prinzip Hoffnung* that Jonas proposes his *Prinzip Verantwortung*, as *The Imperative of Responsibility* is called in German. In what follows, we will more or less follow Jonas' way of procedure and look at the following three points: 1) the identification of the utopian ideal, 2) its feasibility or practical achievability, and 3) its desirability.

The Identification of the Utopian Ideal

According to Jonas' explanation of the basic tenets of Marxist anthropology, human beings are fundamentally good. If they become immoral and corrupt, this is due to the obtaining social conditions. By changing these conditions, we can create a new human being—a superman—who

122. Jonas, *Imperative of Responsibility*, 201.

123. McKenny, *To Relieve the Human Condition*, 48.

124. Bloch, *The Principle of Hope*.

will live in a society of justice and peace, which will be a classless society and "will bring with it the good man."[125] Marx, as Jonas points out, was more interested in the way of getting to these new conditions—the revolution—than in the manner in which a post-revolutionary society would function: "Everything is . . . focused on the revolution and its stages, that is, on the process of *bringing about* the new order. Contrary to the earlier utopias, it is the *coming*, not the *being*, of utopia about which Marxism has something to say." At most it can describe this "being" negatively, that is, "the evils of class society will have *disappeared*."[126] That the post-revolutionary state cannot be described positively is entirely consistent with Marxist ideology. The revolution will change the human situation so radically that, as Jonas points out, it amounts to a new creation, which "will be so novel in the *conditions* of human existence, and so liberating for the hitherto inhibited human potential, that no comparison with the past still applies and all former history pales to prehistory. . . . How this 'true' being will concretely look—the *content* of the vita nova due to reveal itself after the rebirth—of that, no description can be ventured from the state of untrue being in which we are still submerged."[127] In any case, in the society of the "true man," human beings will be good, and organizing them in society will be very easy, so that in this context Lenin once spoke about a washerwoman who would be able to administer the entire affairs of the state after work.[128]

For Marx, the main problem that corrupts human persons is the condition of lack and want. Hatred, envy, violence and all other evils derive from the fact that people have to compete over scarce resources. Thus, the way to remedy all social evil is to create a society of superabundance where everyone gives "according to his abilities" and receives "according to his need."[129] As Robert Spaemann points out, strictly speaking the point here is not so much to create a just society, but rather to create a society in which justice will no longer be necessary—simply because all goods are available in superabundance.[130] Now to Marx's mind a revolu-

125. Jonas, *Imperative of Responsibility*, 158.

126. Ibid., 179.

127. Ibid., 177.

128. Cf. ibid., 117.

129. Marx, *Critique of the Gotha Programme*, 321; cf. Jonas, *Imperative of Responsibility*, 194.

130. Cf. Spaemann, *Basic Moral Concepts*, 36–37: "The peculiar thing about Karl Marx's vision of the future is that it is not concerned with justice, but with setting up

tion was necessary to achieve this purpose in order to see to the equal distribution of wealth, which for him was possible only if the means of production were made the property of all. Here of course communism differs from capitalism, which allows for private ownership of the means of production and which is more tolerant with regards to a fairly unequal distribution of the wealth produced. Where communism and capitalism are in perfect agreement, however, is in the prior presupposition that what matters most is to produce wealth and that the way of doing so is modern technology with its industrial means of production.[131] Marx was convinced that with the technological and industrial revolution we had tools and machineries at our disposal that were so powerful and productive that they could produce a superabundance for all. What had to be changed by way of revolution were the social structures that prevented the equal distribution of all these goods produced by the new machinery. In the post-revolutionary society it was modern technology that was meant to produce such superabundance that would fundamentally alter the human condition and hence bring about the new human being. The way to utopia for Marx, as Jonas reads him, leads by way of modern technology: "It is not by chance that socialism appeared with the beginnings of the machine age and that its scientific accreditation by Marx was based on the condition of capitalism created thereby. . . . Only this condition made socialization seem *worthwhile* . . . Only modern technology makes possible such an increase of the social product that its just (equal) distribution does not result in general poverty."[132]

As little as Marx says about life in this post-revolutionary society of superabundance, one thing is clear: its idea is that people will be free from the necessities which previously forced them to labor. The key

_____

a situation where there is no more need for justice, a situation where there is abundance and everyone can just help themselves without there being any need to pay. The production of goods in this state of abundance should take up so little time that there would be no need for the criteria of justice even in the distribution of work time. This situation would be called 'communism' and in it the principle 'to each according to his needs' would hold sway."

131. Cf. Jonas, *Imperative of Responsibility*, 144: "Wherever socialism came to power, industrialization was the hallmark of its actual and resolute politics. . . . Marxism . . . is no less dedicated to the Baconian idea than its capitalist rival, with which it competes here. To equal and finally surpass it in the fruits to be earned from technology was everywhere the law of its will. In short, Marxism is by origin heir to the Baconian revolution and in its own view its rightful executor, a better one (meaning: more efficient) than capitalism has been."

132. Ibid., 143–44.

aspect of Marx's utopia is freedom from necessity and hence freedom from labor: "The realm of freedom actually begins only where labour which is determined by necessity and mundane considerations ceases."[133] Incidentally, Hannah Arendt is quick to point out that here we touch on a point of flagrant contradiction in Marx's thought, the kinds of which "rarely occur in second-rate writers," but which in the case of great authors "lead into the very center of their work." The contradiction to her mind lies in the fact that "in all stages of his work he defines man as an *animal laborans* and then leads him into a society in which this greatest and most human power is no longer necessary."[134] For Marx, those living in a post-revolutionary communist society are truly free from the necessities of life, and thus they can do whatever they feel like; they are people who "do this today and that tomorrow, who hunt in the morning, go fishing in the afternoon, raise cattle in the evening, are critics after dinner, as they see fit, without for that matter ever becoming hunters, fisherman, shepherds or critics."[135] In short, as Jonas puts it, the necessary condition of utopianism—as exemplified here in Marxism—is "material *plenty* for satisfying the needs of all," and the formal essence of it is *leisure.*[136] Already Hannah Arendt remarked that what remains for people to do after they have been freed from labor and the necessity to make a living, is to engage "in those strictly private and essentially worldless activities that we now call 'hobbies.'"[137] Twenty years later Hans Jonas will take up this clue by his friend and colleague and make an examination of "the hobby" a central point of his critique of Marxist utopia.

The Feasibility or Practical Achievability of the Utopian Ideal

Before moving to an internal criticism of the ideal of leisure and the related notion of the "hobby," Jonas looks at the necessary material conditions of utopianism, arguing that already from this perspective it is inherently impractical. The production of the material plenty required

---

133. Marx, *Capital: A New Abridgment*, 470; cf. Jonas, *Imperative of Responsibility*, 193.

134. Arendt, *Human Condition*, 104–5.

135. Marx, *Deutsche Ideologie*, 22 and 373, cited in Arendt, *Human Condition*, 118n65.

136. Jonas, *Imperative of Responsibility*, 186.

137. Arendt, *Human Condition*, 118.

by the utopian ideal would have a great impact on nature: "The question to be asked here is not how much *man* is still able to do . . . but how much of it *nature* can stand. That there *are* limits of tolerance nobody now doubts, and in our context the question is only whether 'utopia' lies inside or outside of them."[138] At the end, for Jonas the answer is quite clear: utopia lies outside the limits of what nature can bear. In part this is due simply to the energy problem. The reason that this utopian ideal of a labor-free society has at all become thinkable in the first place is of course the industrial revolution—the invention of machines that labor for us. The industrial revolution in turn had its own conditions, among which was the discovery of large quantities of fossil fuels.[139] Oil is the life-blood that keeps our industrial societies running. The creation of a utopian world society—and nothing less could be the aim, since utopianism is by its nature pervasive—would require the solution of the energy problem. Now the energy problem has many aspects, some of which are open to being solved by human ingenuity, while others, to Jonas' mind, inherently do not admit of any solution, for which reason a utopian society will always remain impossible.

The section on the energy problem is noteworthy, especially in the German version of Jonas' book, which is more extensive here. While the English edition essentially contains the same thoughts, it is much more succinct. Still today this passage is highly relevant, despite the fact that it was written some thirty years ago, which in the technological age usually amounts to an eternity. The energy problem as Jonas presents it is twofold. First, there is a difficulty on the supply side: fossil fuels are going to be exhausted in the foreseeable future and feasible alternatives are either shortcoming (solar power, wind energy, water power) or have their own grave difficulties connected with them (nuclear power, with

138. Jonas, *Imperative of Responsibility*, 188.

139. Cf. Hannah Arendt's comments in Arendt, *Human Condition*, 148n9: "One of the important material conditions of the industrial revolution was the extinction of the forests and the discovery of coal as a substitute for wood. The solution which R. H. Barrow (in his *Slavery in the Roman Empire* [1928]) proposed to 'the well-known puzzle in the study of the economic history of the ancient world that industry developed up to a certain point, but stopped short of making progress which might have been expected,' is quite interesting and rather convincing in this connection. He maintains that the only factor that 'hindered the application of machinery to industry [was] . . . the absence of cheap and good fuel, . . . no abundant supply of coal [being] close at hand' (123)."

its radioactive waste and the risk of grave accidents).[140] As we are keenly aware of today, one of the greatest problems with fossil fuels—apart from the fact that they are nonrenewable—is the CO_2 that is released by their combustion and that as a greenhouse gas is most likely co-responsible for global warming.[141] For Jonas the most promising future source of energy is nuclear fusion, which produces very little nuclear waste, excludes the risk of uncontrolled reactions, and does not emit any polluting gases.[142] As it was thirty years ago, still today this form of producing energy is only a future possibility, not a current reality, although some hopeful steps have been made toward it, and the first experimental nuclear fusion reactor (ITER) is currently being built in Cadarache, France by an international consortium.[143]

Could nuclear fusion then serve to solve the energy problem once and for all? Jonas is clear on this point: it could not, because there is not only the supply side of the problem but also a difficulty connected to the very *use* of energy. What for him is an insurmountable problem inherent to the human use of energy is what he calls the problem of thermal pollution. Every use of energy results in the emission of heat, which raises "the question of the admissible upper limit of heating up the closed box of our terrestrial environment." For him, this difficulty cannot be solved by any technological trick, as it is simply consequent to one of the most basic laws of physics: "There is no appeal from the second law of thermodynamics."[144] In other words, it is impossible "to have the *use* of energy without the thermal consequences" for the same reason that it is impossible to construct a *perpetuum mobile*.[145] Therefore, even if we were

140. Cf. Jonas, *Imperative of Responsibility*, 190.

141. Cf. ibid.

142. Cf. ibid.

143. Cf. www.iter.org (last accessed: November 15, 2012).
There is, however, one rather recent development that is not reflected in Jonas' presentation, and that is the reappraisal of solar power, a source of energy which Jonas mentions without giving it much appreciation. Recent calculations have suggested that the United States, for instance, could provide for their entire energy demand by setting up solar power stations in its southwestern deserts, using existing technologies. For instance, there is the Kramer Junction solar power plant in California's Mojave desert, which has been successfully operating since 1989. For all this see: Zweibel et al., "Solar Grand Plan," 64–73. The authors in fact argue that given the potentials of solar power, the energy problem is not one of technological art but of political will.

144. Jonas, *Imperative of Responsibility*, 191.

145. Ibid., 190–91.

blessed with a basically inexhaustible, clean, and safe source of energy, we would still have to use it responsibly for the simple fact that its use produces heat. A global utopian society with its unfettered use of energy would simply overheat.

To Jonas' mind what is needed therefore is modesty in the goals of technological progress. This call to caution is issued by the imperative of responsibility. Renunciation will be required of us, and we must certainly not aim at utopia, which "insofar as harnessed to material plenty [is] *the* immodest goal par excellence." This is so "not only because, if ever attained, it could not last, but more so because already the road in that direction leads to disaster."[146]

The Desirability of the Utopian Ideal

Having, at least to his mind, demonstrated that the utopian ideal is inherently impracticable, Jonas argues that it is nonetheless important to show that this ideal, even if it were possible, is in itself undesirable. For him the need for this demonstration derives from two facts. First, he mentions that under one condition a utopian state would be physically possible after all, namely "with a sufficiently low or lowered number of humans! If one does not shrink from the monstrous measures required for that, an island of the blessed may well be established for the chosen few on the corpses of countless discards."[147] Evidently, here we see another major problem of utopianism, apart from its impracticability: it can lead to fanaticism that deems every means justified to achieve its end. In Jonas' words, "one must not underrate the lengths to which a highest good absolutely believed in may seduce its self-appointed trustees."[148] If the authentic human being comes to exist only in the post-revolutionary utopian society, then, human beings as they exist today are not truly human and can be sacrificed for the sake of their future true being.[149]

146. Ibid., 191.

147. Ibid., 192.

148. Ibid.

149. In this context Pope Benedict XVI severely criticizes the Marxist theory of impoverishment, which says that in order to create a future authentic humanity, one should precisely *not* offer any assistance to the working poor, which would only serve to prolong unjust social structures. Rather, the aim should be that, under the pressure of poverty, there will finally be a revolution which will cause these unjust structures to collapse. This is indeed one of the ways in which utopia sacrifices present humanity

Another reason why Jonas continues his critique of utopianism is that even as disillusioned about the possibility of achieving its ideals, it can still be a powerful idea by leaving people simply disappointed. They may resign themselves to the impossibility of attaining the ideal without thereby necessarily beginning to see anything good in the world as it happens to be right now. Their disappointment may still be of great social consequence. The negation of the present implied in utopianism can very well survive its affirmation of a better future once the latter has proved to be impossible: "Beyond the dangers of uncritical faith there are those of disappointed faith: the dangers of despair when merely the external unattainability and not also the intrinsic error of the ideal comes to be grasped. For the ideal, if true, condemns every other state to being unworthy of man; and it is not good to go into the soberness of renunciation with hate for that with which one has to live thereafter."[150]

In his discussion of the desirability of the utopian ideal, Jonas asks the decisive question: Granted that in such a world we would be free *from* the necessity of labor, what is it that we would be free *for?* What would take the place of labor, filling the empty space left behind by it? A world freed from necessity would be a world of freedom for *what?* As we have mentioned, Jonas' argument here is that Marx does not give much of an answer, since he assumes that once the social conditions are such as to produce "the new human being," society will simply organize itself. In addition, what would we, "the old humans," have to say about the way that "the new humans," whom we are about to create, should lead their life? They will have leisure, but how to fill this leisure is what Marx wants to leave up to them to decide. Ultimately, of course, this response simply dodges the question and is hardly intellectually satisfying. Hence, Jonas seeks another interlocutor, turning to the Marxist philosopher Ernst Bloch. Like Marx, Bloch argues that in a communist society people will be free from labor and that they will have leisure, but then he goes a step further, qualifying this leisure as "active," in some ways evoking Aristotle's

---

to the future one. In response Benedict argues, "Seen in this way, charity is rejected and attacked as a means of preserving the *status quo*. What we have here, though, is really an inhuman philosophy. People of the present are sacrificed to the *moloch* of the future—a future whose effective realization is at best doubtful. One does not make the world more human by refusing to act humanely here and now" (Benedict XVI, *Deus Caritas Est*, n. 31).

150. Jonas, *Imperative of Responsibility*, 192.

idea of happiness that is not simply a state but rather an activity.[151] As to the concrete content of this active leisure, Bloch makes two proposals: the hobby and the cultivation of interpersonal relationships.[152] Jonas submits both of these to a penetrating critique.

THE HOBBY

Even in an industrial, post-revolutionary society of superabundance a few true professions will remain: engineers and programmers who design the machines, inventors, doctors, teachers. However, the vast majority of people will be liberated, or, as Jonas puts it, excluded from any serious activity.[153] Under such conditions, people will want to work not because there is any need for its results, but rather because they sincerely desire to do so, so that "the need for work as such becomes a therapeutic rather than economic issue."[154] The demand for work will exceed its availability. To this problem Ernst Bloch proposes a solution: the hobby as "profession" or ultimate vocation of the human person.[155] Previously, when they were still under the sway of necessity, people enjoyed practicing their hobbies on the side. Now they can enjoy practicing them full-time.

Jonas takes issue with this proposal on three accounts. In his view it would lead to a loss of spontaneity, a loss of freedom and a loss of reality. As to the loss of spontaneity, people enjoy their hobbies precisely because they can practice them at their whim, as they like it. They can take it or leave it. It is a characteristic mark of the hobby to be an activity practiced on the side, spontaneously without necessity, an activity which needs its contrast with one's main occupation. The moment the hobby becomes one's main occupation, it will lose precisely the characteristics that defined it as a hobby: its spontaneity and its counter-balancing relief function. As Jonas puts it, "I doubt that even the machinist whom circumstances rather than love have gotten into the job would wish to collect butterflies all the time."[156] Indeed, from the fact that one and the

151. Cf. ibid., 195.

152. The question of interpersonal relationships Jonas only treats in the original German edition of his work.

153. Cf. ibid., 194.

154. Ibid.

155. Cf. ibid., 195.

156. Ibid., 196.

same practice, such as gardening, for instance, can be a hobby for one and a profession for another, we see that it belongs to the very idea of hobby to be a side-activity.

Secondly, Jonas sees a threat to freedom under Bloch's scenario. Thinking the idea to its end, he argues that in a really existing utopian society everyone would *need* to have a hobby, simply to avoid the risk of criminality, drug abuse, or whatever other bad idea people can come up with when they have too much time in their hands. The duty to have a hobby would have to be imposed by the government, which also would have to see to the right distribution and assignment of hobbies. After all, hobbies are expensive, and they will have to be paid for by public resources, since private resources would have been abolished. Besides, they do not produce anything except "a hole in the public budget,"[157] and thus cannot pay for themselves. As a result, "utopia which donates the leisure must also govern it with a paternalistic hand."[158]

The highest price to be paid however is the ultimate loss of reality, which is the third point of Jonas' criticism. Inasmuch as hobbies are un-productive, or at least insofar as their product—should there be any after all—does not matter, they are not serious practices. This is the one defin-ing characteristic that remains unaltered even when they are turned into a "profession." They will lose their spontaneity but they will not thereby gain any seriousness. The hobby as a profession robs the realm of human activities of that seriousness that alone is compatible with human dignity. The utopian society is one in which everything that is done could just as well be left undone or be done poorly without consequence, because—as a leisure activity—it did not matter in the first place. As a result, "the ghostliness of unreality descends on the whole make-belief activity, and with it an unimaginable *taedium vitae* whose first victim must be the pleasure in even the self-chosen hobby."[159] For Jonas, this world of pre-tense and make-belief does not take persons seriously, to whose dignity it belongs to prove themselves in the seriousness of life, to confront reality and to stand their ground. He points out that the basic fallacy underlying Marxist—and for that matter any other—utopianism is to think that the realm of freedom begins only where the realm of necessity ends, while, in fact, freedom, in order to be at all, "lives in pitting itself against necessity.

157. Jonas, *Das Prinzip Verantwortung*, 360 (translation my own).

158. Jonas, *Imperative of Responsibility*, 197.

159. Ibid.

Separated from it, freedom loses its object and becomes as empty as force without resistance. Empty freedom, like empty force, cancels itself."[160]

If then the hobby will not do as content of Bloch's active leisure, how about the cultivation of interpersonal relationships? Jonas agrees that even in a world in which all the basic necessities of life are taken care of, there still remains the realm of human relationships with its own dramatic nature and seriousness, with its "love and hatred, cold and warmth," with its hopes and disappointments.[161] But what would it mean to make *it* the object of one's leisure, that is in this context to dedicate one's main time and effort to it? To the extent that the state—as we have already mentioned—would need to take a high interest in its citizens' object of leisure, making sure they stay busy lest the devil find work for their idle hands, the privacy of human relations could no longer be maintained. The state, in any case, would seek to intervene in this previously most private of realms: the realm of the family and friendship, which would hardly support such interference without major damage.[162]

The most fundamental problem as Jonas sees it, however, is yet another: interpersonal relationships and friendships, he points out, are always *about* something. Here he echoes Aristotle's idea of the common good as the basis of friendship. In addition, we can hear a resonance of the notion of "world" that Hannah Arendt put forth. The world, for Arendt, is the metaphorical space that at the same time separates and unites people in their common concern for it.[163] Jonas himself formulates it this way, "One must have world to be 'oneself' for oneself—how much more to be for others! Friendship is a covenantal allegiance for some-

160. Ibid., 198.

161. Jonas, *Das Prinzip Verantwortung*, 366. Unfortunately, Jonas did not include the section on interhuman relationships as content of leisure in the English edition of his work. It is noticeable how generally in the later sections of the book, the English version tends to become much more succinct than the German version, which may indicate that Jonas, who oversaw the translation, felt under pressure to finish the work. It is true that Jonas' overall argument can also live without this particular section, so that one can see why under a presumed pressure he would have just left it out. However, this does not mean that it is not significant in itself.

162. Cf. ibid., 367.

163. Cf. Arendt, *Human Condition*, 52.

thing and against something *in the world*; it is ultimately rooted 'in the common concern,' and the other is treasured because he treasures similar things."[164] But if friendship is indeed based on this common good and this common concern, then it can only thrive if there is something in the world that really matters. This is even the case with marriage, which is a "community of care in maintaining oneself in life-long necessity; and the pleasure of love appears before the background of the seriousness of a shared reality."[165] It will be difficult to maintain a friendship, then, when all seriousness is gone—in a society of superabundance where there is nothing left to be fought for or against. Without the daily occupations of life, without anything that human relationships can be about, when they themselves are made the direct object of leisure, they can only become "pathological, parasitical, and cannibalistic."[166] Hence, human relationships as object of leisure will not do either. The utopian ideal is nothing that is in itself desirable.

Concluding Remarks on Utopia: The "True Human Being" Is Always There Already

As we have said at the beginning of this section, Jonas emphasizes that his critique of Marxist utopianism is also "implicitly a critique of technology in the anticipation of its extreme possibilities."[167] This, we mentioned, was so for at least two reasons. First, Marxist utopianism is built on the hypothesis of technological progress. The revolution is only one aspect of Marxism and regards the equal distribution of superabundant goods. That the goods to be distributed are in fact superabundant is not itself owed to the revolution, but rather to industrial and technological progress. Secondly, however, the alliance between Marxism and technology is so close that we must wonder whether there is not a utopian tendency inherent to the idea of technological progress itself. What does progress strive toward? Is it not the ideal of a utopian society of universal welfare, freed from necessity in which people can live a life of ease? And even more, apart from some concrete goals toward which technological progress strives, in many ways progress has itself become the ideal of "the real

---

164. Jonas, *Das Prinzip Verantwortung*, 368 (translation my own).

165. Ibid., 369 (translation my own).

166. Ibid. (translation my own).

167. Jonas, *Imperative of Responsibility*, 201.

vocation of man": as "an indefinite, self-validating advance to mankind's major goal, claiming in its pursuit man's ultimate effort and concern," its ideal has become its own moving forward.[168]

In this sense, for Jonas the most dangerous problem that Marxism and the idea of technological progress have in common is the one defining characteristic of any kind of utopianism: to devalue the present for the sake of the future to the point of sacrificing or at least risking present humanity for a future enhanced or more authentic one; in short, they share what he calls an "ontology of 'not yet.'"[169] To this erroneous position he wants to oppose with all force "the plain truth . . . that genuine man is always already there and was there throughout known history: in his heights and his depths, his greatness and wretchedness . . . in all the *ambiguity* that is inseparable from his humanity."[170] This ambiguity, for Jonas, is inherent to the very nature of the human being, and to try to eliminate it would amount to abolishing the person.[171] He warns us that "the really unambiguous man of utopia can only be the flattened, behaviorally conditioned homunculus of futuristic psychological engineering. This is today one of the things we have reason to *fear* of the future. *Hope* we should, quite contrary to the utopian hope, that in future, too, every contentment will breed its discontent, every having its desire, every resting its unrest, every liberty its temptation—even every happiness its unhappiness."[172]

Jonas ends his book with a plea that there is something we must never do: we must not, with the magic tools of our technology, risk or

168. Ibid., 4.

169. Ibid., 200. Again, while Jonas discusses Bloch's Marxism here, its two central points—that true humanity is not yet and that technology will serve us to bring it about—are characteristic of the "Baconian project" of technological progress as such. Hence, McKenny suggests that some of those working in the field of biomedicine—the present-day cutting edge of technological progress—may well be guided by utopian ideals when he writes, "In addition to Bloch's (and Bacon's) vision of a humanity freed from the ceaseless round of toil, we may add the visions of biomedical utopians who seek to enhance human traits, forestall or reverse the processes of aging, or alter human behavior" (McKenny, *To Relieve the Human Condition*, 46).

170. Jonas, *Imperative of Responsibility*, 200.

171. McKenny aptly summarizes Jonas' position in this way: "Human nature, normatively considered, just *is* the living out of the polarities that constitute it, and therefore is always in process but never still to be realized. To seek to realize our own fleeting conceptions is to terminate the process in favor of one arbitrary distortion of it" (McKenny, *To Relieve the Human Condition*, 69).

172. Jonas, *Imperative of Responsibility*, 201.

sacrifice the image of humanity as we know it for the sake of a future humanity. We are custodians of the image called to pastor and care for the inheritance that has been entrusted to us. Karl-Otto Apel tends to agree with Jonas that "what matters in the present situation is to save the *existence*—hence the *survival*—and the *undamaged image of the essence*—hence the *dignity*—of the human being from the dangers that lie in the mere continuation of technological progress and hence in the mere advance of the current process of industrialization."[173] Nonetheless he raises an important objection, wondering whether we may not need some ongoing technological progress simply to preserve humanity, which, by nature is both inventive and in need of inventiveness: "One can . . . ask oneself whether the existence and dignity of the human being can be saved at all by *merely preserving the now existing state of affairs*. More precisely, is the nature of man and of his long since technologically and socio-culturally transformed environment not so constituted that without a *regulative idea of technological and social progress* it cannot be preserved?"[174]

Here we can respond that for Jonas technology and its progress is not anything bad in itself nor are the obtaining social conditions sacred to him. He is not proposing our return to a society of hunters and gatherers or the permanent solidification of the status quo. In fact, quite along the lines of Apel's concerns, at some point Jonas speaks about a "progress in technology [that] is necessary," namely "simply for the correction of its own effects." But then immediately he goes on to say that this "does not cancel out the advice to be modest."[175] What he warns against is being so immodest as to make progress an end in itself—to see in it the vocation of the human being—and to sacrifice what has been entrusted to us today for the sake of the future. He would not object to seeking some technological progress and to doing whatever else is necessary to *preserve* our inheritance. What he calls immoral is to run inestimable risks for the sake of *improving* it: "The great risks of technology . . . are not undertaken to preserve what exists or alleviate what is unbearable, but rather to continually improve what has already been achieved, in other

---

173. Apel, "Problem of a Macroethic," 7.

174. Ibid.

175. Jonas, *Technik, Medizin und Ethik*, 71 (translation my own).

words, for *progress,* which at its most ambitious aims at bringing about an earthly paradise."[176]

Jonas' emphasis on the preservation—not of existing social conditions but of the nature of the human being—may nonetheless invite criticism. Thus, Lucien Sève wonders whether respecting future human beings should not involve doing "everything in our power, so that they dispose of the maximum *autonomy,* and consequently to pass on to them the greatest number of means possible?"[177] Besides, while he grants to Jonas that some of our choices can be irresponsible, he contests that responsibility always has to be negative, i.e., we may have the responsibility to help our descendants become better humans: "[Jonas] absolutely refuses to admit . . . that by our responsible choices we can on the contrary favor for them an authentically human existence, or, in other words . . . let *progress* be made in humanization."[178]

It may be to Sève's surprise, but most likely Jonas would entirely concur with the aims that he proposes, i.e., to help our descendants become autonomous and make "progress" in their humanization. As to the increase in autonomy, Jonas' whole argument was that our technological toys have an ambivalent nature and to some extent always also make us their servants and not only their masters. With the presence of a wide variety of tools also comes the necessity of their use, and here our descendants will then be impaired in their autonomy.[179] This does not say that these tools are bad; it only shows that "the greatest number of means possible" does not necessarily amount to the greatest autonomy. To help our descendants become autonomous, we rather have to be prudent—and modest. When it comes to making progress in humanization, my guess is that Jonas would be all for it—only that he would understand this to mean progress in being truly *moral* agents, which first implies that we will abstain from manipulating our descendants with our technology and secondly, that we will take up the task of educating them, where "we also impart, at least do not foreclose, the means for revision, and certainly leave the inherited nature what it is."[180]

176. Jonas, *Imperative of Responsibility,* 36.

177. Sève, *Pour une critique de la raison bioéthique,* 181 (translation my own).

178. Ibid. (translation my own).

179. This is of course even more so if the technological tools are made to bear directly on our descendants by interfering with their bodily heritage as would be the case with possible future genetic manipulation.

180. Jonas, *Philosophical Essays,* 153n12.

One problem, however, must certainly be raised in this context, and both Apel's and Sève's remarks point to it: It may at times be difficult to distinguish between cases where we legitimately seek to preserve what exists or attempt to remove unbearable evil and where we do things that are completely optional, i.e., where we act for progress as an end in itself, committing acts for which Jonas would reserve the name "frivolous."[181] Much of what he says about responsibility, caution, and modesty seems to depend on this distinction between the dutiful preservation of what has been given to us and its optional enhancement. However, this distinction, as we will see in the final chapter, is itself not entirely unproblematic. Here we would simply like to suggest that at least in theory it can be upheld and that, although there may be many cases where it is difficult neatly to differentiate the two, there will also be many situations where the difference is quite clear to common sense.

In any case, Jonas' emphasis on preservation does not imply an abnegation of action or a prohibition against changing obtaining states of affairs. On the contrary, Jonas argues that preserving the image is a *task,* and that it is a task for which we need *hope,* which is the presupposition of every action insofar as it is the conviction that one can do something effectively.[182] The worst thing would be a fatalism that resigns itself to the situation, saying that one cannot make a difference anymore anyway, that things will go their way in any case.[183] In addition, when he speaks about *fear*—which might suggest timidity—in no way has he in mind a fear of the paralyzing kind that is afraid of what is to come in the future, but rather a fear *for* the precious object that is threatened at present.[184] At the heart of the ethics of responsibility which Jonas proposes is a "veneration for the image of man, turning into trembling concern for its vulnerability."[185] According to him, we need to cultivate a sense of trembling at the mere imagination of what may possibly happen to the human being. From this trembling Jonas hopes that we may recover the sense of reverence for the image which we are called to care for and pastor.[186] Presumably, what he has in mind here is the *aidos* of the Greeks

181. Cf. Jonas, *Imperative of Responsibility,* 77.

182. Cf. Jonas, *Das Prinzip Verantwortung,* 391 and Jonas, *Imperative of Responsibility,* 201–2.

183. Cf. Jonas, "Fatalismus wäre Todsünde," 455–56.

184. Cf. Jonas, *Das Prinzip Verantwortung,* 391.

185. Jonas, *Imperative of Responsibility,* 201.

186. On another occasion Jonas, speaking in theological language, points out that

who knew it as a secular yet holy fear of offending something sacred.[187] Cultivating this reverence for the image of man will help us to protect its integrity. Jonas concludes by pointing out that this task, while it is not utopian, will nonetheless not be easy: "To preserve the integrity of his essence, which implies that of his natural environment; to save this trust unstunted through the perils of the times, mostly the perils of his own overmighty deeds—this is not a utopian goal, but not so very modest a task of responsibility for the future of man on earth."[188]

An Appreciation of *The Imperative of Responsibility*

Jonas' book *The Imperative of Responsibility* is truly valuable already for its analysis of the ambiguity of the technological age in which technological progress has taken on a life of its own, unfettered by any whence or whither. That technology to a large extent serves us only to solve problems which it first helped to create has become somewhat of a commonplace for us today, who have long since made the experience that just as for every problem there is a solution, so to every solution there is a problem. But we must remember that the book was written thirty years ago. Of course, in the midst of the Cold War, very few people doubted the possibility that humanity could face destruction by means of technology—namely the nuclear bomb. Yet perhaps here one should not so much speak of the *use* of technology but rather of its *abuse*. To have pointed out that even the proper use of technology, which does not aim at annihilating humankind

---

this reverence implies mostly a negative duty, namely "we must not try to fixate man in any image of our own definition and thereby cut off the as yet unrevealed promises of the image of God. We have not been authorized, so Jewish piety would say, to be makers of a new image, nor can we claim the wisdom and knowledge to arrogate that role" (Jonas, *Philosophical Essays*, 181).

187. Cf. Jonas, *Das Prinzip Verantwortung*, 392–93: "We must learn again to cultivate a reverence and a trembling, that they may protect us from the wrong tracks of our power. . . . The paradox of our situation consists in the fact that we have to regain our lost reverence from our trembling, the positive from the imagined negative: the reverence for what the human person was and is from the trembling at the face of what he might become. . . . Reverence alone, by revealing to us something 'sacred'—something that must not under any condition be injured (and this can be seen also without positive religion)—can protect us from defiling the present for the sake of the future, from wanting to buy the latter at the price of the former" (translation my own).

188. Jonas, *Imperative of Responsibility*, 202.

but is rather meant to be to its benefit, could unleash similar powers of destruction was one of Jonas' true and authentic contributions.

Jonas' practical advice as to what we should do is for the most part very useful and valuable. Trying to train our imagination in order to regain a sense of the sacredness and dignity of the image of the human person as this image has been passed on to us, using a heuristics of fear to learn what is at stake, refraining from ever putting the being and the good being of humanity on the stakes—all these are points well-taken. In fact, most people would probably find it hard to disagree with his new categorical imperative: "Act so that the effects of your action are compatible with the permanence of genuine human life."[189]

More controversial, however, are two other issues. One of these we have already briefly alluded to. It concerns the relation between the existence and the good existence of humanity as Jonas sees it. Could we ever risk the good existence of humankind to secure its mere existence or should it be the other way around? Here Jonas opens himself up to some criticism and serious concern. The second issue regards precisely his central notion of responsibility, which raises a good number of questions, to which we shall turn in due course.

The Relation between the Sheer Existence and the Good Existence of Humanity

We may wonder whether Gerald McKenny does not indeed point to a certain irony in Jonas' thought. As Jonas analyzes the ambiguity of the Baconian project of gaining power over nature, he explains that the problem with the Baconian project is that it is ultimately too successful. Human persons who now dominate nature are in danger of losing the power over their power. It is as if in its success the Baconian power over nature went on a kind of autopilot, henceforth dictating us the way to proceed. The human person's power over nature has obtained power over human beings themselves. In Jonas' words, "The power has become self-acting, while its promise has turned into threat, its prospect of salvation into apocalypse."[190] What Jonas seeks—and here is the irony—is to find ways of gaining power over this power: "Power over power is required now before the halt is called by catastrophe itself—the power to overcome that

189. Ibid., 11.
190. Ibid., 141.

impotence over against the self-feeding compulsion of power to its progressive exercise."[191] What we need is to dominate domination. With this
suggestion, however, he still seems to be thinking within the very Baconi-
nan framework that he criticizes. As McKenny puts it, "Jonas ultimately
must understand responsibility itself in quasi-Baconian terms as gaining
power over power or regaining control over what now threatens to control us."[192] The issue continues to be one of power, even if this time—for
Jonas—we are speaking of "a third-degree power," i.e., a power over the
human powers over the powers of nature.[193] Jonas, in our view rightly, indeed questions one of the fundamental premises of the Baconian project,
that is, the idea that human beings are confronted with a hostile nature
that imposes on them the alternative of seeking domination over it or
allowing themselves to be dominated by it in a struggle that is fought
to the death.[194] When Jonas suggests, however, that now that nature has
proven herself the weaker duelist, we need to dominate our own powers
of domination, he does not leave the battlefield but simply exchanges one
opponent for another.

Whatever we may think of McKenny's observation—and indeed it
should be hard to present Jonas in the breadth of his thought as a full-
fledged Baconian—there is at least one aspect of his thinking where
certain Baconian tendencies may in fact be verified and indeed lead to
fairly grave consequences, particularly in the one rather disastrous section of what is otherwise a truly commendable book. In his search for
the ways to dominate our power—to tame the forces we ourselves are
unleashing by our technological capabilities—Jonas seriously wonders
whether a dictatorship or even a totalitarian regime might not be much
better suited to solve our ecological problems than a democracy.[195] And

---

191. Ibid.

192. McKenny, *To Relieve the Human Condition*, 74.

193. Cf. Jonas, *Imperative of Responsibility*, 141–42.

194. Cf. Jonas, *Phenomenon of Life*, 193: "Not only is man's relation to nature one
of power, but nature herself is conceived in terms of power. Thus it is a question of
either ruling or being ruled; and to be ruled by a nature not noble or kindred or wise
means slavery and hence misery. The exercise of man's inherent right is therefore also
the response to a basic and continuous emergency: the emergency of a contest decreed
by man's condition."

195. Cf. Jonas, *Imperative of Responsibility*, 150–51: "In the preceding deliberations
it was tacitly assumed that, in the coming severity of a politics of responsible abnegation, democracy . . . is at least temporarily unsuited, and our present comparative
weighing is, reluctantly, between different forms of 'tyranny.'"

if people will not change their ways if one tells them the truth, then a well-taken lie will have to do.[196]

It seems that here, for a brief unguarded moment, Jonas falls prey to the very danger that he argued was inherent to utopianism: if there is any earthly human goal that we set as absolute, and if this goal be as noble as peace on earth and justice for all, the danger will be that these goals will be taken to justify any means for achieving them. If world peace were indeed our highest good, then nothing should hinder us from establishing a society in which people are hooked on *soma* and bred according to social needs. Few however will find Huxley's *Brave New World* an appealing place, and some, upon reading the novel, may indeed feel grateful that we still live in a world where war—as great an evil as it is—is still possible. Indeed, there are certain great goods—among them human freedom in this fallen world—that can only be had at the price of the *possibility* of this evil.

Perhaps quite unawares, Jonas himself has set up a mundane good as the highest that in all its humility could still function just the same way as more ambitious, utopian ideals, namely the survival of humanity, so that he opens himself up to the charge of "a *utopianism of responsibility.*"[197] It seems that to Jonas this good is so high that to ensure it we could also institute dictatorships or resort to lies. The question is whether Jonas is consistent here with the rest of his thought. Karl-Otto Apel actually thinks this to be the case, arguing that from his premises Jonas is not able to deduce principles of justice toward currently living human beings.[198] For Apel, Jonas' imperative only prescribes humanity's ongoing existence, and the ways to achieve that could even be highly immoral, for which

Hence it is not without reason that Pier Paolo Portinaro calls Jonas both "a prophet and a tyrant" (cf. Portinaro, "Il profeta e il tiranno," 100–111).

196. Cf. Jonas, *Imperative of Responsibility*, 149: "Perhaps this dangerous game of mass deception (Plato's 'noble lie') is all that politics will eventually have to offer: to give effect to the principle of fear under the mask of the principle of hope."

197. Cf. Apel, "Problem of a Macroethics," 27.

198. Cf. Apel, "Ecological Crisis," 240–41: "Even if one grants Jonas' metaphysical premises for the sake of the argument, it seems to me to be completely impossible, in principle, from these premises to derive not only the duty of securing the preconditions of humankind's further existence (without depravation!), but also—simultaneously—the duty of respecting the *equal rights of all living human beings* with regard to the future existence of humankind. For Jonas' attempt at deriving novel versions of a 'categorical imperative' from his first imperative shows, as far as I can see, that, on his premises, he cannot derive a satisfactory substitute for Kant's *universalization principle of justice.*"

reason Jonas' approach is wanting: "His proposed formulations . . . can only prescribe that life's, and especially humankind's, existence ought to continue; but this does not preclude that even a racist solution of the problem (e.g., a solution at the cost of a starvation of the Third World peoples) could be considered as a fulfillment of the demand."[199]

It seems to us that Apel has in fact touched a neuralgic point here, and Jonas' speculations about dictatorships and the abolition of political freedoms would appear to confirm him in this. In fact, on a different occasion Jonas explicitly states that "all moral standards for individual or group behavior, even demands for individual sacrifice of life, are premised on the continued existence of human life."[200] While one of his central points of critique against utopianism—which strives for the supreme good—is precisely that it can instill a fanaticism that may not even "shrink from the monstrous measures" required for actively lowering the number of humans living on earth,[201] when the bare survival of humanity itself is at stake, he suggests that what we now call ethical may have to be suspended: "Averting disaster takes precedence over everything else, including pursuit of the good, and suspends otherwise inviolable prohibitions and rules."[202] Indeed, Jonas claims that in a true "lifeboat situation," where either some go overboard or all will drown, "all rules cease to apply."[203] Humanity "can again be adorned by ethical conduct" only "after the storm has been weathered."[204] Thus, he can imagine the scenario of what he calls a "moral pause," the moral inference of which, however is precisely that "we must never allow a lifeboat situation for humanity to arise."[205]

To our mind there is indeed a grave problem here with Jonas' thought. It is certainly wise to say that we must avoid a lifeboat situation. But even in such a situation, no one has the right to kill another, no one may become a murderer. Whether some freely and heroically renounce their lives for the sake of the others, whether they draw lots, or whether they all drown—all these scenarios are better than one in which some are

199. Apel, "Ecological Crisis," 240–41.

200. Jonas, "Toward a Philosophy of Technology," 43.

201. Jonas, *Imperative of Responsibility*, 192.

202. Jonas, "Toward a Philosophy of Technology," 43.

203. Jonas and Scodel, "An Interview with Professor Hans Jonas," 367.

204. Jonas, "Toward a Philosophy of Technology," 43.

205. Ibid.

reduced to the state of wild beasts, violently throwing others overboard. Life is not the highest good. As Socrates reminds us, it is better to suffer from injustice than to commit it.[206] It is better to drown than to become a murderer. And this must even be applied to humanity as such and not only to individual personal lives.

It may seem here that we are simply affirming the opposite of what Jonas wants to say. However, there is some way in which even Jonas, on his own terms, would have to agree or at least admit that there is some inconsistency in his thought. In his new categorical imperative, he speaks about the "permanence of genuine human life."[207] We mentioned before that Jonas left the relation between humanity's *sheer* existence and its *good* existence somewhat unspecified. But we also recall that for Jonas humanity ought to be because responsibility ought to be, or, in other words, because of their capacity for morality. A humanity as described by Huxley no longer exists in a humane way, as it is no longer capable of responsibility. The goal that Jonas proposes with his new categorical imperative is to secure the survival of humanity *as responsible beings,* which hence includes the preservation of the conditions for such responsibility, among which is not existence alone but also openness to truth and freedom. It seems at least odd to suggest that there could ever be a situation in which—in order to preserve human beings as responsible beings—we should act toward them irresponsibly, in fact morally repulsively, or that we should enslave people or lie to them in order to safeguard their capacity for freedom and truth. As Stanley Hauerwas aptly points out, even though we need to survive in order to treasure things, there are things we treasure more than our survival: "Unless human beings survive, it is impossible for them to pursue and achieve other values; existence is the indispensable condition for the realization of values. At the same time, it may be doubted that people want to exist merely to exist; they live in order to pursue goods other than that of sheer survival itself."[208] The words of the Roman poet Juvenal come to mind here, who directs them to the individual, but which may well apply to the whole human race: "Consider it the greatest of crimes to prefer survival to honor and, out of love of physical life, to lose the very reason for living."[209]

206. Cf. Plato, *Gorgias*, 474b.

207. Jonas, *Imperative of Responsibility*, 11.

208. Hauerwas, *Truthfulness and Tragedy*, 122.

209. Juvenal, *Satirae*, VIII, 83–84 as cited by Pope John Paul II, *Veritatis Splendor*, n. 94.

The existence of future humanity is not something to be *constructed* or achieved, cost what it may—including lies and tyranny. Rather, it is something to be *preserved* by abstaining precisely from any acts that could compromise it. As the invention of the nuclear bomb, for instance, puts the *mere* existence of human beings at stake so does tyranny with their *good* existence. We should notice also that when we look at Jonas' imperative itself, we find that indeed it is either worded negatively or—in a positive reformulation—at least has a negative sense: What is sought is compatibility of one's actions with the future good existence of humanity. He does not actually say, "Act in a way to bring it about cost what it may." Hence, Jonas' imperative does not necessarily imply that for the sake of ensuring humanity's *sheer* existence we must under certain conditions resort to acts that compromise its *good* existence, understood as a moral existence capable of truth and freedom. In fact, his imperative may actually preclude such acts. As Spaemann puts it, "the first and unconditioned responsibility toward *every* human being is a negative one: to avoid and to renounce influencing them in such a way that does not respect them as persons."[210] There do seem to be things we never ought to do, things that are morally impossible for us, even if by our omission we risk putting the future of humanity at stake.[211] To ensure the permanence of *genuine* human life, then, there are some acts we will always have to omit, such as throwing people off the lifeboat.

The Notion of Responsibility

RESPONSIBILITY AND NONRECIPROCITY

The difficulties we have just raised—Jonas' authoritarian tendencies and the difficulties he has in accounting for relationships of justice among contemporaries—are in some ways the result of his emphasis on the non-reciprocity of responsibility and the interpretation of the paradigms that he chooses for it: the relation between a parent and a newborn and the

---

210. Spaemann, *Happiness and Benevolence*, 182.

211. Cf. Spaemann, "Wer hat wofür Verantwortung?" 237, who argues that we cannot be held responsible for the consequences of our omitting immoral acts and that here the moral boundary—the moral "I-cannot"—should be regarded in the same way as the physical boundary. "We know that no one is held responsible for the consequences of omitting something that was physically impossible for him to do. But there are also moral impossibilities" (translation my own).

relation between a head of state and his or her country. Those for whom some have responsibility seem entirely passive, and the way that Jonas describes the head of state's responsibility for the nation makes us think of authoritarianism and dictatorship.[212] Is it true that responsibility is always nonreciprocal? And how are we best to interpret Jonas' paradigms of responsibility?

For Jan Schmidt there is no doubt that the dimension of nonreciprocity is in fact one of the central points of Jonas' ethics. The concept of responsibility serves to bring this dimension to the point. In fact, "*nonreciprocity* is the formal core of the condition for the possibility of responsibility."[213] On the same page Schmidt, apparently intending to rephrase the same idea in different words, says, "There can only be responsibility where there is an asymmetry of power. . . . Asymmetry of power is a condition of an ethically relevant situation."[214] But is it true that asymmetry of power and nonreciprocity are really the same, or could we not take these two expressions as pointing to an important distinction? It seems in fact undeniable that responsibility presupposes power, that is, that I can be responsible for something only if in some ways it has come under my influence to promote it or to harm it. It is however doubtful that—at least insofar as responsibility for human beings is concerned—this relationship of responsibility, asymmetrical in power though it be, could ever be entirely nonreciprocal. Nonreciprocity suggests not only an uneven balance of power but a complete one-way street, implying that the one who has responsibility is an agent while the one falling under that responsibility is a mere patient, i.e., an object that is entirely passive. It is true that in a relationship of responsibility one person has responsibility for another and not the other way around. However, it is also true, as Jonas himself admits, that this situation could change at any time, depending as it is, simply on the contingent factors of power distribution and special needs.[215] Thus Karl-Otto Apel remarks, "Jonas' typical examples,

212. Lucien Sève speaks of Jonas' "defiance of democracy" and his "elitism and paternalism," which are the "only two political expressions that Hans Jonas can ultimately give to an ethics that had started out on a good path" (Sève, *Pour une critique de la raison bioéthique*, 183; translation my own).

213. Schmidt, "Aktualität der Ethik von Hans Jonas," 555 (translation my own).

214. Ibid. (translation my own).

215. Cf. Jonas, *Imperative of Responsibility*, 94: "Among natural brothers responsibility arises only when one of them falls on evil days or otherwise needs special help." It is hard to see why Jonas thinks that such circumstances are so exceptional and not

the responsibility of parents for their children and the responsibility of the statesman for the weal and woe of the citizens entrusted to him, fail to show that responsibility is *not a relationship of reciprocity*. They rather show that the *fundamental responsibility of human beings for one another* is a *potential* relationship that becomes *actual* only in accordance with a real advantage of power."[216]

In fact, human beings are never merely passive; they are also always *agents,* so that when we take responsibility for humans, we take responsibility for beings that are themselves responsible. Therefore, they are never simply the *objects* of our responsibility but rather its *addressees*, who can ask an account from us for what we did.[217] For this reason Robert Spaemann writes, "Helping a person is . . . something different from helping an animal. It always means: helping a helper. We have to justify our help *to the person* himself. We are not to circumvent their freedom as a subject in order to help them as a natural being."[218] But the moment I remember that the other whom I am helping is a *responsible being,* there is some *reciprocity*, namely an interaction from responsible being to responsible being that requires my accountability, even though there is a momentarily uneven distribution of power.

This even holds true for our responsibility for newborns or for future, not-yet existing generations. Although they are not actually able to exercise responsibility, the way we take responsibility for them must be very different from the way we deal with other living beings. At any stage of their development, human beings are persons[219] and as such the

---

rather more or less the rule—which seems more consistent with the way things really are.

216. Apel, "Problem of a Macroethic," 19.

217. Cf. Spaemann, *Happiness and Benevolence*, 181.

218. Ibid., 182.

219. For this, see Robert Spaemann's very helpful arguments summarized in the last chapter of his book *Persons*. Thus, he convincingly argues that there are no "potential persons"; "something" will never turn into "someone": "There are, in fact, no potential persons. Persons possess capacities, i.e., potentialities, and so persons may develop. But nothing develops *into* a person. You don't become some-one from being some-thing. If personality were a condition of affairs, it could arise bit by bit. But a person is a someone situated *in* this or that condition; the condition is always a predicate of the person, the person always presupposed by the condition. The person is not the result of modification; it simply 'presents itself', like substance in Aristotle. The person is substance, because the person is the mode in which a human being exists" (Spaemann, *Persons*, 246).

kind of beings that are capable of responsibility, even if they may not yet be able to exercise it in actuality. For this capacity to become actual, however, it is true that first others have to take responsibility for them. As Spaemann puts it, "The young infant becomes a speaking being, i.e., one who is capable of self-determination, only after others have taken over responsibility for it."[220] Whatever form the exercise of our responsibility for newborns or infants may take, however, it will always have to be in a way that we can anticipate our having to give them an account for it. We must already help them as one helps helpers, even though they cannot actually help themselves nor anyone else yet—just as we speak to them as one speaks to speakers, even though they cannot actually speak or understand yet. But when in speech we address them as persons—as living beings "who have the word,"[221]—we are not pretending. It is precisely by being addressed in speech that they will learn to speak, that they will become conscious of themselves and finally also capable of taking responsibility for themselves and others. Just as newborns already are the speaking-kind-of-beings before they can *actually* speak, so they already are responsible-kind-of-beings even before they can *actually* take responsibility. Therefore, when we take responsibility for them, the relation is never entirely nonreciprocal but always from responsible being to responsible being, even though of course the power is asymmetrical.

This is also the case with future generations. We are responsible for them as to beings who are themselves responsible. In this sense, there is reciprocity, even though in terms of power, there is an uneven balance. Our power over them is much greater than their power over us, albeit one cannot even say that they have *no* power over us at all. The ancient Roman practice of the *damnatio memoriae*—the blotting out of the memory—for instance, is one form in which a younger generation can exercise power over an older one, even if that older generation has already naturally and peacefully passed away. Besides, as Aristotle points out, even after their death it is no matter of indifference to persons how their descendants are faring and whether their enterprises succeed or fail.[222] Inasmuch as it has

220. Spaemann, *Happiness and Benevolence*, 183.

221. Cf. Aristotle, *Politics*, I, 1, 9. For a profound reflection on the relation between "the word" or language and the spiritual nature of the human being, see Ebner, *Das Wort und die geistlichen Realitäten*.

222. Cf. Aristotle, *Nicomachean Ethics*, I, 10: "Both evil and good are thought to exist for a dead man, as much as for one who is alive but not aware of them; e.g. honours and dishonours and the good or bad fortunes of children and in general of descendants."

power over the older generation's descendants and projects, a younger generation also has power over the older generation.

GLOBAL RESPONSIBILITY

Besides, a question arises about the subject of responsibility. Jonas himself admits that in our new technological age the agents are "not you or I: it is the aggregate, not the individual doer or deed that matters here." Therefore, the "imperatives of a new sort" that our situation requires are primarily directed to public policy.[223] As Günter Maschke points out, in such a situation, then, it is impossible for individuals to assume responsibility, as they receive no account of the consequences of their concrete actions.[224] Where conglomerates or abstract entities like "science," "the economy," or "the government" act on a global scale, individuals will feel powerless and be tempted to resign themselves to the recognition that they cannot make a difference anyway. Hence, we may ask whether at the end of the day Jonas is not in some sense incapacitating individuals by taking away their responsibility and attributing it to "global players." Is Jonas' ethics of responsibility ultimately making people irresponsible, leading them to resignation by confronting them with a universal responsibility?[225] Spaemann, for instance, seems to agree with Jonas that there is such a thing as a responsibility for the world in its entirety, since "the dimensions of human activity have reached such a range that at least its unintended side-effects can affect the Earth as a whole." However, he immediately goes on to say that "it is not easy to be clear about what kind

223. Jonas, *Imperative of Responsibility*, 9.

224. Cf. Maschke, "Das Pflügen des Meeres und die Begründung der Ethik," cited in Apel, "Problem of a Macroethic," 11.

225. For the problem of a universal responsibility that is not mediated by an *ordo amoris*, see Spaemann, "Wer hat wofür Verantwortung?"
For how a wrongly-intended sense of responsibility or "compassion" can wreak havoc particularly in the medical field, see Hauerwas, *Dispatches*, 165: "Modern medicine has had its task changed from care to cure in the name of compassion—a killing compassion. For example, the recent discussion of doctor-assisted death, or what perhaps should be called doctor-assisted suicide, surely must be seen in this context. Unable to cure those who are dying, we then think it is the compassionate alternative to help them to their death. Euthanasia thus becomes but the other side of the medical and technological imperative to keep alive at all cost." He then goes on to formulate matters in more general terms: "When compassion becomes the overriding virtue, linked with liberal political practice, it cannot help but be destructive" (Hauerwas, *Dispatches*, 165–66).

of responsibility this is and who is its bearer."[226] For Spaemann, responsibility in its ordinary sense "derives from the situations in which we find ourselves, i.e., from moral relationships. Moral relationships are: friendship, marriage, the relationship between parents and their children, between physicians and their patients, between teachers and their students, between colleagues, and so on."[227] And in fact, the examples for responsibility that Jonas indicates are precisely taken from the context of such moral relationships, particularly the parent-child relationship, but even the one between the head of state and the country can be interpreted in this way. We must ask ourselves then about the relationship between this global responsibility for the world as a whole and the responsibility that stems from concrete moral relationships. Can the former ever be interpreted in terms of the latter? Who are its subjects and can the danger of the individual's resignation in front of an impossible global task be avoided?

Our main aim here is not so much to answer these very difficult questions but rather to show that Jonas' concept of responsibility raises them. Nonetheless, one promising way of interpreting global responsibility in a manner that may avoid the dangers of resignation and grounds it in concrete moral relationships is understanding it in terms of *co-responsibility*, an idea proposed by Karl-Otto Apel and others. Thus, Marianna Gensabella Furnari pointedly observes that usually children have *two* parents, so that already Jonas' "parental paradigm is dual and, as such, is not the paradigm of responsibility, but of co-responsibility."[228] Proceeding along the lines of this thought, however, she wonders whether perhaps a better paradigm of our responsibility for the earth and for the future of humanity would not be the co-responsibility of brothers for a common project. These brothers would then be responsible individually for each other, depending on contingent factors, and together and collectively responsible for the common enterprise. To her mind, in this way one can bridge the gap between individual responsibility, which, in Spaemann's terms is rooted in moral relationships, and collective responsibility for the world as a whole: "The paradigm of fraternity, intended as responsibility of one toward the other of men who are *here* and *now*, and as co-responsibility of all for the common enterprise (the survival of

226. Spaemann, "Wer hat wofür Verantwortung?" 229 (translation my own).
227. Ibid., 226 (translation my own).
228. Gensabella Furnari, "Ontology of Temporality," 149.

authentic human life on Earth) appears to be the most suitable model for the passage from individual responsibility to collective responsibility."[229]

Apel, too, presupposes the fraternal paradigm, arguing that it is precisely the notion of collective responsibility that can respond to the "objection concerning the impotence of the individual in the face of the unforeseeable consequences of our collective actions." To avoid the danger of resignation, we need to "realize that the decisive point is by no means that the individual *alone* ought to assume responsibility for the future."[230] Lacking power, an individual usually does not have personal responsibility for the way a multinational conglomerate treats global resources. The moment we speak of the "responsibility" of that same global player, we notice that we fall short of personal responsibility. In most cases, not even the CEO has enough power to be called personally responsible for the direction of his or her company. We speak here of corporate responsibility, which is ultimately the responsibility of no one. Now Apel argues that a responsibility that is at the same time personal and global has to be collective. It is neither individual nor corporate, and it is not abstract but concrete: "Indeed, the point is only that even as he reads the morning newspaper the individual should think how he, according to his own competencies and abilities, can *take part in organizing collective responsibility*."[231] Individuals can do something precisely by coordinating their activities with others. Participating in "practically relevant discourses," individuals can organize their collective responsibility, which is a process that takes place "on innumerable organizational levels, both institutional and informal—from the level of legislation to that of establishing a kindergarten or to that of a pensioner composing a letter to the editor of a newspaper."[232] Following this fraternal paradigm, then, responsibility will remain personal—it is rooted in the context of moral relationships, such as brotherhood or friendship that is in pursuit of a common endeavor—and at the same time transcend the individual—it is collective responsibility precisely *for* that common project. If Arendt is right and coordinated activity results in power,[233] then people taking

229. Ibid., 153.

230. Apel, "Problem of a Macroethic," 27.

231. Ibid.

232. Ibid.

233. Cf. the chapters "Power and the Space of Appearance" and "Unpredictability and the Power of Promise" in Arendt, *Human Condition*, 199–207 and 243–47 respectively.

co-responsibility will in fact be able to effect changes; theirs will be a true responsibility, and they will have no reason for resignation.

RESPONSIBILITY AND LOVE

The question was whether an individual could ever be the subject of responsibility in Jonas' sense, since it is a responsibility for an object—nature as a whole, the future of humanity—that seems to exceed an individual's power. But if the only subjects of responsibility were corporate persons such as nations or companies, then Jonas would have effectively abolished the whole idea of human responsibility inasmuch as this idea presupposes some concrete human actors. To speak of a corporation's or parliament's responsibility is not nonsense, but something much less concrete than to talk about John's and Ann's responsibility. Apel's idea of collective responsibility is a way in which individuals could meaningfully understand themselves to be responsible even for objects that by far exceed their individual power without having to abnegate their initiative to some impersonal corporate entity. Furthermore, by making the parent-child relationship one of his paradigms for responsibility, Jonas implies that he does not mean to restrict the application of this concept to the global level, essentially limiting it to corporate "global players." For him, our—collective—responsibility for the future of the environment and of humanity is thus not essentially different from a mother's and father's responsibility for their children.

If this is so, and with whatever problems this understanding raises, we may wonder whether there is not a simple answer as to why—as Jonas notes—the feeling of responsibility has been so conspicuously absent in previous ethical theory.[234] The response would simply be that responsibility *has* actually always been at the center of morality, only that previously it has gone by a different name, and that is: love or benevolence. Thus, Wolfgang Kuhlmann writes, "To me it seems overall justified to consider this principle [of responsibility] to be a specific from of the principle of '*benevolence.*' Here I understand benevolence to be an attitude or mind-set of an acting subject toward his environment, by which he first and foremost (*prima facie*) attempts to realize the good for his environment or for the relevant aspect of it—and this for the sake of his environment (or its relevant aspect)."[235] Jan Schmidt, too, sees an analogy between Jo-

234. Cf. Jonas, *Imperative of Responsibility*, 87.

235. Kuhlmann, "'Prinzip Verantwortung' versus Diskursethik," 288 (translation

nas' idea of responsibility and benevolence. He points out that "already prior to Jonas, William K. Frankena has analogously spoken of a 'principle of benevolence.'"[236] Schmidt then goes on to refer to William James' very relevant observation, "Take any demand, however slight, which any creature, however weak, may make. Ought it not, for its own sole sake, to be satisfied? If not, prove why not."[237] But this is precisely benevolence or love, namely, "To be-out-for that which is beneficial to the other, that is, that which fulfills the other's being-out-for."[238] It means to perceive the teleology of a being and to take it up into one's own teleology. Therefore, according to Spaemann for a benevolent person "to see a beetle lying on its back struggling and to turn it over are one and the same,"[239] while not to do so, as James has pointed out, would require a justification. This indeed seems to be very much what Jonas has in mind when he speaks of responsibility as a sentiment that is provoked in us by the perception of a vulnerable being together with our conviction that we have the means of coming to its aid. To see the helpless baby and to see that he ought to be, to perceive his cry and to perceive his claim on our will to come to his aid, are one and the same thing. As Jonas says, I can mute this call by the process of scientific abstraction and tell myself that this being is just a lump of cells, but then it is no longer the baby I see.[240] We may therefore

---

my own).

236. Schmidt, "Aktualität der Ethik von Hans Jonas," 553 (translation my own). Cf. Frankena, *Ethics* (1963), 37: "We have a prima facie obligation to maximize the balance of good over evil, if and only if we have a *prior* prima facie obligation to do good and prevent harm. I shall call this prior principle the *principle of benevolence.*" It should be said, however, that for Frankena this principle of benevolence is part of a utilitarian approach, which to our mind Jonas would not be likely to endorse. Besides, in the second edition of his work, Frankena decided that he preferred rather to speak of the "principle of beneficence": "We have a prima facie obligation to maximize the balance of good over evil only if we have a prior obligation to do good and to prevent harm. I shall call this prior principle the *principle of beneficence.* The reason I call it the principle of *beneficence* and not the principle of *benevolence* is to underline the fact that it asks us actually to do good and not evil, not merely to want or will to do so" (Frankena, *Ethics* [1973], 45).

237. James, "Moral Philosopher," 249.

238. Spaemann, *Happiness and Benevolence*, 97.

239. Ibid., 178.

240. Cf. Jonas, *Imperative of Responsibility*, 131: "The theoretical rigorist may ask: What is really and objectively 'there' is a conglomeration of cells, which are conglomerations of molecules with their physicochemical transactions, which as such *plus the conditions of their continuation* can be known. . . . But is it the infant who is seen here? He does not enter at all into the mathematical physicist's view, which purposely

wonder whether Jonas' principle of responsibility is not a specific form of the principle of benevolence,[241] which is precisely our openness to letting ourselves be affected by the claim that the being of reality makes on our will.

The main problem with Jonas' global responsibility is thus the same as the problem of a universal love. The problem is that it asks too much of us as finite beings, which is why a truly universal responsibility can only be a negative one: to abstain from doing harm.[242] In this negative sense, all of the biosphere along with future human generations can become the object of my responsibility to the extent—and only to the extent—that I am able to abstain from certain compromising actions. When it comes to changing more complex social and even global structures, however, the subject of responsibility has to become more and more collective—I have to take responsibility together with my friends in accordance with what is really in our common power. But even collective responsibility, the moment it aims at more than not doing harm, cannot be truly universal, since not even collectively people have universal power. Collective responsibility will have to be broken down according to an *ordo amoris*, the proper ordering of our loves and responsibilities that takes account of our finitude. As Spaemann puts it, "Our responsibility for and to humans is grounded in the claim of every human being to be taken by every other rational being not just as an object, but as a self-being. That this claim cannot at every moment be made good by everyone toward everyone is due to the finitude of humans, and this finds its rational expression in that which we have named, following the tradition, *ordo amoris*."[243]

If we are right with our suggestion that Jonas' notion of responsibility is really a type of benevolence or love, then we can rephrase the upshot of his argument for the "ought-to-be" of humanity. Jonas' "Humanity ought to be because responsibility ought to be" will then mean,

---

confines itself to an exceedingly filtered residue of his otherwise screened-off reality."

241. Thus, Kuhlmann argues, "It is a *special* form of the principle of benevolence because Jonas emphasizes (1) that the resulting obligation concerns only the avoidance of evils, (2) that the environment in question is explicitly not restricted to one's vicinity, (3) that despite the great importance of the affections the principle is rather meant to have the character of a rational principle" (Kuhlmann, "'Prinzip Verantwortung' versus Diskursethik," 288–89; translation my own).

242. Cf. Spaemann, *Happiness and Benevolence*, 182: "The first and unconditioned responsibility toward *every* human being is a negative one: to avoid and to renounce influencing them in such a way that does not respect them as persons."

243. Ibid.

"Humanity ought to be because love ought to be." A world in which there are not only purposes, but in which there are beings capable of making other beings' purposes their own is a world that is infinitely better than one in which each being only pursues its own purposes or one in which there is no purpose at all. We may thus conjecture that implicitly in Jonas love is not only the highest form that life takes on—which was the conclusion of our discussion of his *The Phenomenon of Life*—but that love is also the highest reason for why humanity ought to be—which may stand as the conclusion of our discussion of his *The Imperative of Responsibility.*

3

Jürgen Habermas and Genetic Enhancement

Jonas' Contribution to the Contemporary Debate

IN THIS FINAL SECTION of our book, we would like to argue for the continued relevance of Jonas' thought for the contemporary debate on biotechnology. Very concretely we will do so by looking at its role in Jürgen Habermas' booklet *The Future of Human Nature*,[1] which argues against a liberal eugenics and which already from the considerable feedback that it has received—to say nothing of its generally well-laid out argument—is qualified to count as an important contribution to present-day reflections on these issues.[2] Habermas draws on Jonas at two central steps of his line of reasoning, while perhaps the most interesting congruence lies in the *structure* of the argument itself, which has surprisingly much in common

1. Habermas, *Future of Human Nature*.

2. Cf. for instance: Birnbacher, "Habermas' ehrgeiziges Beweisziel," 121–26; Dumitru Nalin, "Liberté de procréation et manipulation génétique," 31–54; Dini, "Natura umana e biotecnologia," 33–56; Fenton, "Liberal Eugenics and Human Nature," 35–42; Junker-Kenny, "Genetic Enhancement as Care," 1–17; Kuhlmann, "Wider die Verdinglichung des Menschen," 55; Mendieta, "Habermas on Human Cloning," 721–43; Moss, "Contra Habermas ," 139–49; Neubach, *Das Selbstseinkönnen eingebettet in der Gattungsethik*; Prusak, "Rethinking 'Liberal Eugenics'," 31–42; Siep, "Moral und Gattungsethik," 111–20; Song, "Knowing There Is No God," 191–211; Spaemann, "Habermas über Bioethik," 105–9; Viano, "Antiche ragioni per nuove paure," 277–96.

with Jonas' *The Imperative of Responsibility,* even though Habermas of course emphatically does not share any of Jonas' metaphysical commitments.[3] Both Jonas and Habermas argue from transcendental, i.e., necessary conditions for morality. Though they differ in where they locate the core of the human being's moral capacity—for Jonas it is the capacity to take responsibility, for Habermas it is the ability to enter into a discourse among equals—both are concerned precisely about *the conditions of its possibility.*[4] In other words, both argue that with some of our technological advancements—particularly in the field of biotechnology—humanity's very capacity to lead a moral existence could be at stake.[5] Hence, both of them go beyond saying that this or that procedure is morally good or morally bad, moral or immoral for this or that reason. The danger they see is that we could construct a being that has lost his or her capacity for being either moral or immoral.

But why could some biotechnologies bring with them the risk of abolishing the human person as a moral being? Here, both Jonas and Habermas actually agree: the danger is that with these technologies we may *impose our own image on future generations.* For Jonas we lack the wisdom to do so; we deal irreverently with the image of the human person as we have received it; we run intolerable risks, and we set ourselves

3. Thus, Alessandro Dini comments on the relation between Jonas and Habermas, "Jonas and Habermas move within very different theoretical perspectives.... However one can trace convergences and affinities between the positions of Jonas and Habermas. To confirm his assertions, Habermas cites a passage from *Philosophical Essays: From Ancient Creed to Technological Man* in which Jonas emphasizes the difference between interventions on inanimate matter and interventions on animated matter. But it is above all in the conception of the human being that their positions find points of contact" (Dini, "Natura umana e biotecnologia," 48; translation my own).

4. That this concern is common to Jonas' approach and to discourse ethics in general was seen by Wolfgang Kuhlmann. Just as discourse ethics is concerned with preserving the conditions of the possibility of morality, so is Jonas' ethics of responsibility: "The highest value in the ethics of responsibility, the value from which moral demands are derived, is the capacity of living beings to have ends, which is a capacity that in the human being is still surpassed in the capacity for morality (and that is evidently a condition for efforts aiming at justice). The first moral obligation according to Jonas can thus also be formulated in this way: it is the obligation to ensure that morality continues to be possible" (Kuhlmann, "'Prinzip Verantwortung' versus Diskursethik," 296–97; translation my own).

5. This is also the great concern that Francis Fukuyama expresses in his *Our Post-human Future,* 102: "What is ultimately at stake with biotechnology is not just some utilitarian cost-benefit calculus concerning future medical technologies, but the very grounding of the human moral sense."

up as a master generation that decides about the fate of future generations. It is this latter point that Habermas explicitly picks up from Jonas, developing it further in the context of his own discourse ethics, showing how by imposing our image on our descendants, we no longer relate to them as equals but introduce a new form of domination, never known before, into our relationship to them. In this way we may very well inhibit their own sense of freedom and their ability to see themselves as the undivided authors of their own lives.[6] This capacity in turn is essential, because for Habermas people are socialized to become moral subjects precisely in communicative contexts where persons essentially relate to each other as equals.

The other point where Habermas more or less explicitly relies on Jonas is in the context of dealing with what is perhaps the most significant objection that can be leveled against his argument, namely the claim that by enhancing our children by biotechnological means—such as biogenetic manipulation—we do not do anything that is essentially different from what we do when we educate and socialize them. Here both Jonas and Habermas argue that in the case of education, children can—at least at a later point—take a critical distance from the history of their socialization, i.e., they can accept or reject it. With direct genetic interventions on the body, which serve to form part of the person's very identity, no such reflexive stance is possible.

Habermas' Life and Thought

Before going into a more detailed discussion of Habermas' *The Future of Human Nature,* we will first provide some important background about his person and thought. Habermas was born in Düsseldorf, Germany, in 1929 and has thus lived through World War II in his childhood years. From 1949 to 1954 he studied philosophy, history, psychology, literature, and economics at the universities of Göttingen, Zürich, and Bonn.[7] In 1954 he completed his doctoral dissertation on Friedrich Schelling at the University of Bonn. Moving to Frankfurt in 1956, he became assistant to Theodor Adorno at the Institute for Social Research, where he began

6. Cf. Habermas, *Future of Human Nature,* 63.

7. For the following biographical and bibliographical data we mainly rely on Andrew Edgar, *The Philosophy of Habermas,* and James Gordon Finlayson, *Habermas. A Very Short Introduction.*

to work on his *Habilitation*. As he had differences of opinion with Max Horkheimer, who was supposed to be his moderator, he left in 1958 to go to Marburg, where he finished his work in 1961. Directed by Wolfgang Abendroth,[8] Habermas' second major academic research project examines the social development of England, France, the USA, and Germany from the eighteenth to the early twentieth century, a time in which these countries experience a transformation from being feudal and monarchic societies to becoming bourgeois and finally mass societies. The book was translated into English only in 1989 as *The Structural Transformation of the Public Sphere: An Inquiry into a Category of Bourgeois Society.*[9] In 1961 Habermas received a call to the University of Heidelberg, which was issued on the initiative of Hans-Georg Gadamer with whom he was to engage in a famous debate on the value of hermeneutics a few years later.[10] When in 1964 he was offered the chair of philosophy and sociology as the successor to Horkheimer, he went back to Frankfurt. Seven years later he moved to Starnberg close to Munich, where he co-directed the Max-Planck-Institute. In 1981, after some disputes with his co-workers, he returned to the Institute for Social Research in Frankfurt, where he taught philosophy until his retirement in 1994. He can be considered one of the most influential living philosophers in continental Europe. His influence also extends into the Anglo-Saxon world, as he makes frequent visits to the United States, giving guest lectures, and many of his interlocutors are in fact American or British.

Habermas is a prolific writer, who, together with Karl-Otto Apel, belongs to the second generation of the Frankfurt School of Critical Theory, and his name is often mentioned in one breath with those of his teachers and colleagues Theodor Adorno, Max Horkheimer, and Herbert Marcuse, who represent the Frankfurt School's "first generation." These latter started out hopeful that in their critical theory they had an instrument with which to scrutinize and test philosophical or social theories for ideological conditionings and unwarranted prejudices in order to arrive at conclusions that are strictly speaking reasonable. However, particularly after World War II, they became more pessimistic as to what reason can achieve by itself. Thus, one of Horkheimer's noted lines is "Knowing there

8. Cf. Stirk, *Critical Theory, Politics, and Society*, 26.

9. Habermas, *Strukturwandel der Öffentlichkeit*; English: Habermas, *Structural Transformation of the Public Sphere*.

10. Cf. Palmer, "Habermas versus Gadamer?" 487–500 and Madison, "Critical Theory and Hermeneutics," 463–85.

is no God, it [critical theory] nevertheless believes in him."[11] In their famous book *Dialectic of Enlightenment*, Adorno and Horkheimer show the entanglements of an enlightened reason that, following Kant, dares to know for and by itself. At some point, enlightened reason, with all its achievements, reaches a dead-end, being unable, for instance, to find any basic argument against murder.[12] In his evaluation of human reason, Habermas, in contrast, is more optimistic than his mentors. For him, the project of Modernity is merely unfinished, but not surpassed.[13] Reason is a universal faculty, which works the same in different times and cultures, and we can really come to know moral rightness. The confidence he has in reason (in contrast to a postmodern defeatism) and the cognitivist stance he takes with regards to morality are perhaps among the reasons why he was able to have a fruitful discussion with the then-Cardinal Ratzinger on the foundations of public morality in Munich in 2004.[14]

His *magnum opus* is his two-volume *Theory of Communicative Action*,[15] in which he argues for the intersubjective nature of human rationality and distinguishes between action that aims at understanding or consensus—communicative action—and action that aims at results—strategic or instrumental action.[16] The moment two or more people live together, they will somehow have to organize and arrange the way they do so. For Habermas, language plays a crucial role in this. As we have to deal with conflicts or agree on common goals, we use language to communicate about these issues and arrive at solutions that everyone can agree on or can at least in some way freely consent to. This is communicative

11. Horkheimer, "Kritische Theorie und Theologie," 508, cited by Habermas, *Future of Human Nature*, 113.

12. Horkheimer and Adorno, *Dialectic of Enlightenment*, 93.

13. Cf. for example: Habermas, "Modernity: An Unfinished Project," 38–55.

14. Cf. Habermas and Ratzinger, *The Dialectics of Secularization*.

15. Habermas, *Theory of Communicative Action*, vol. 1: *Reason and the Rationalization of Society*, and vol. 2: *Lifeworld and System: A Critique of Functionalist Reason*.

16. Cf. the concise summary that Habermas gives of his theory of communicative action in his *Moral Consciousness and Communicative Action*, 58: "I call interactions *communicative* when the participants coordinate their plans of action consensually, with the agreement reached at any point being evaluated in terms of the intersubjective recognition of validity claims. . . . Further, I distinguish between communicative and strategic action. Whereas in strategic action one actor seeks to *influence* the behavior of another by means of the threat of sanctions or the prospect of gratification in order to *cause* the interaction to continue as the first actor desires, in communicative action one actor seeks *rationally* to *motivate* another by relying on the illocutionary binding/bonding effect (*Bindungseffect*) of the offer contained in the speech act."

action: to treat the other, with whom I have to come to an agreement, as an equal partner and to seek a solution to problems by means of rational arguments and in all sincerity so as to arrive at a *consensus*. In contrast to communicative action, instrumental action may be directed to nonhuman things, too. If it is directed to human beings, it does not see these as equal partners. While in communicative action the relationship among the partners in the dialogue is of crucial importance and the aim is to reach an agreement on a given issue, instrumental action is completely governed by the goal toward which it strives. Its aim is not consensus but *success*. If one produces a shoe, it does not matter how one does so, as long as the shoe that comes out is of good quality. In terms of human relationships, instrumental action refers to any kind of interaction in which one person makes other persons do something or in some way acts on them without entering into a communicative context with them. This could be done, for instance, by using incentives, by telling them lies, by threatening them, or by using physical violence.

On the basis of his theory of communicative action, Habermas—together with Karl-Otto Apel—elaborated his version of a discourse ethics. His main concern here is to show the possibility of a cognitive ethics in the absence of any metaphysical foundations, to which, to his mind, one can no longer make any appeal. According to Habermas, discourse itself contains moral commitments as its transcendental pragmatic conditions. For instance, on the pain of a performative contradiction, anyone engaging in discourse needs to acknowledge the fundamental freedom of access to the debate and the equality of all discourse partners; besides, the very practice of discourse, in order for it to be possible, presupposes the truthfulness of all participants and the absence of coercion.[17] While these transcendental pragmatic conditions constitute what Habermas at one point called an "ideal speech situation,"[18] i.e., they will never real-

17. Cf. Habermas, *Justification and Application*, 31: "Anyone who seriously engages in argumentation must presuppose that the context of discussion guarantees in principle freedom of access, equal rights to participate, truthfulness on the part of participants, absence of coercion in adopting positions, and so on. If the participants genuinely want to convince one another, they must make the pragmatic assumption that they allow their 'yes' and 'no' responses to be influenced solely by the force of the better argument."

18. Cf. Adams, *Habermas and Theology*, 23, who argues that "Habermas abandoned the ideal speech situation over twenty years ago—in name and in substance." When however Adams goes on explaining what Habermas meant by the "ideal speech situation," Adams is in fact describing what later would become part of Habermas'

istically be achieved in their pure form, they do serve to govern *actual* discourses and not simply imagined ones. This means that Habermas' discourse ethics, even though it is of Kantian inspiration, is not simply a tool with which people by themselves can envision what is to be done, imagining the ideal speech situation in a similar way as they would imagine a maxim that could become a universal law. Rather, discourse ethics indeed calls on existing people actually to communicate and enter into real dialogue and debate.[19]

Apart from being a social theorist and a proponent of discourse ethics, Habermas over the years has very effectively played the role of public intellectual with incisive interventions on topics of the day, published in national German newspapers or magazines, which he then regularly issued in book form. For one of these collections of articles, published as *Time of Transitions,* in which he reflects on German politics after the fall of the Berlin Wall, he, too, just like Jonas, received the Peace Prize of the German Book Trade.[20] At the end of his acceptance speech, given in the wake of the September 11 terrorist attacks, he presents the key points of his thought on some issues of biotechnology,[21] which in the same year

---

transcendental conditions of discourse: "What is the ideal speech situation? It is an expression of the symmetry between partners in dialogue. For Habermas, symmetry is itself an expression of peaceability, and is emblematic of commitment to rational debate. When two parties enter into dialogue, they have a commitment to treat each other not as 'things' in the world, which they manipulate, but as 'subjects' whose freedom to determine their own courses of action is equal" (27). We would thus suggest that Habermas abandoned the "ideal speech situation" only in name rather than in substance. And his motivation may have been that the name was misleading. Inasmuch as it contains the adjective "ideal," it may have evoked the image of a philosopher sitting in his armchair and thinking through practical matters entirely by himself in his utter self-sufficiency. This, as Adams rightly notices, is not what Habermas is all about. Rather, for Habermas "real life comes first, and philosophical abstractions (things like the ideal speech situation) are always subsequent" (40). It is from the experience of actual discourse that Habermas then formulated the transcendental conditions of it.

19. Cf. Habermas, *Moral Consciousness*, 203: "Discourse ethics rejects the monological approach of Kant, who assumed that the individual tests his maxims of action *foro interno* or, as Husserl put it, in the loneliness of his soul. The singularity of Kant's transcendental consciousness simply takes for granted a prior understanding among a plurality of empirical egos; their harmony is preestablished. In discourse ethics it is not. Discourse ethics prefers to view shared understanding about the generalizability of interests as the *result* of an intersubjectively mounted *public discourse.*"

20. Habermas, *Time of Transitions.*

21. Habermas, "Faith and Knowledge," which has been included, very appropriately, in the volume Habermas, *Future of Human Nature.*

he develops in greater detail in a booklet called *The Future of Human Nature*.[22]

The Basic Argument of *The Future of Human Nature*

In *The Future of Human Nature*, Habermas deals with three biotechnological procedures, two of which are current actualities and one of which is a not too remote possibility: preimplantation genetic diagnosis (PGD), stem cell research, and genetic manipulation. PGD is illegal in Germany and stem cell research is restricted to already existing stem lines. On the pragmatic level, the upshot of his argument is a plea to keep things this way. Both practices should be kept illegal or at least continue to be restricted because they lend themselves to positive eugenics. With PGD these eugenic tendencies are inherent to the idea of generating human life conditionally, i.e., making the absence or presence of certain traits the criterion for its continued existence.[23] With stem cell research, the problem, as Habermas sees it, is with an instrumentalizing attitude toward human life which it tends to engender in us, desensitizing us to the respect that is due to life, even though for him this respect is not unconditional at the prenatal stage.[24] Habermas' case against PGD and stem cell research is essentially a slippery slope argument: both, for different reasons, can lead to the establishment of the practice of positive eugenics, and therefore these should be kept in check.[25]

On a side note we should remark how in the context of presenting his case, Habermas introduces the distinction between "the dignity of human life" and "human dignity." According to him "human dignity" refers to the unconditional respect that is due to human persons, who are such because by birth they have entered a socialization process by which they are integrated into "the *public* context of interaction of an intersubjectively shared lifeworld."[26] The "dignity of human life" for Habermas, in contrast, refers to the *certain* respect we feel for human life at other

22. Cf. the German original: Jürgen Habermas, *Die Zukunft der menschlichen Natur. Auf dem Weg zu einer liberalen Eugenik?*

23. Cf. Habermas, *Future of Human Nature*, 21.

24. Cf. ibid., 20.

25. Cf. ibid., 66–74. See also Habermas, "Auf schiefer Ebene," 33, which can essentially count as Habermas' own brief summary of his argument in *Future of Human Nature*.

26. Habermas, *Future of Human Nature*, 34.

stages. To his mind, this respect—respect though it is—is not unconditional but admits of degrees. Thus he refers to our "gradually changing evaluative sentiments and intuitions toward an embryo in the early and middle stages of its development, as compared to a fetus at the later stages."[27] By referring to these presumed intuitions, Habermas means to give support for his distinction between the two kinds of dignity. However, when he appeals to yet another alleged experience or sentiment, he ultimately leads the distinction *ad absurdum*: "It is in this respect that we feel compelled to distinguish between the dignity of human life and human dignity as guaranteed by law to every human person—a distinction which, incidentally, is also echoed in the phenomenology of our highly emotional attitude toward the dead."[28] In his rather strange effort to compare our attitudes toward living prenatal human life to our feelings about the dead, Habermas is led to a somewhat incongruous formulation, which itself reveals that the comparison does not hold water. Thus he speaks about the "due respect toward dead life,"[29] "dead life" of course being a complete oxymoron.[30]

After presenting his slippery slope argument, Habermas dedicates the center piece of his book on showing why positive eugenics, particularly in the form of genetic enhancement, is so problematic. It is here that our own interest lies, since we would like to look at Hans Jonas' part in Habermas' argument, which appears in his discussion of genetic enhancement. For convenience's sake we will for the moment speak as if genetic enhancement were already a feasible procedure, referring to it in the present indicative and not in the more appropriate future conditional.

Some Presuppositions

To understand why Habermas considers genetic enhancement problematic, we first need to recall his theory of communicative action and his discourse ethics, to which we have briefly alluded above. For Habermas

27. Ibid., 32.

28. Ibid., 35–36.

29. Ibid., 36.

30. Cf. also Robert Spaemann's critique of Habermas on this point: "It is telling that Habermas draws a parallel between our dealings with the unborn and our dealings with human beings after their death. Every mother who feels her child moving in her womb will find this counter-intuitive" (Spaemann, "Habermas über Bioethik," 107; translation my own).

the individual becomes a moral subject by entering into a communicative context with others.[31] It is through the medium of language and thus intersubjectively that the persons come to themselves, that they develop self-consciousness and become actors. The "language game of morality," as Habermas calls it,[32] presupposes the fundamental equality of all involved and likewise it requires that all participants can consider themselves the responsible authors of their own life and action.[33]

Second, we need to refer to Habermas' distinction between morality and ethics. Ethics for him is concerned with questions about the good life, which can only be answered within the horizon of a concrete life-form and on the basis of metaphysical or religious assumptions that go unchallenged within that context.[34] In a multicultural, pluralistic and postmetaphysical society, there is, however, little hope of these assumptions being universally shared, for which reason practical philosophy has by and large to remain silent about these matters, although rational discourse about them is not precluded.[35] Morality, on the other hand, deals with questions of justice, which Habermas defines as universalizable

31. Cf. Habermas, *Justification and Application*, 130–31: "Individuation is merely the reverse side of socialization. Only in relations of reciprocal recognition can a person constitute and reproduce his identity. . . . Only in the aggregate of his communicative expressions does a person become identical with himself."

32. Cf. ibid., 43: "The intersubjective character of freedom and practical reason becomes manifest when we analyze the role an expression such as 'respect' assumes in the language game of morality."

33. Cf. Habermas, *Future of Human Nature*, 92: "We can only take part in the moral language game under the idealizing presupposition that each of us carries the sole responsibility for giving ethical shape to his or her own life, and enjoys equal treatment with complete reciprocity of rights and duties. But if eugenic manipulation changes the rules of the language game itself, this act can no longer be criticized according to those rules. Therefore, liberal eugenics provokes the question of how to value morality as a whole."

34. Cf. Habermas, *Erläuterungen zur Diskursethik*, 39: "*Moral questions*, which, under the aspect of universalizable interest or *justice*, can in principle be decided upon in a rational manner, are now to be distinguished from *evaluative questions*, which, under their most universal aspect, present themselves as questions of the *good life* (or of self-realization) and which are open to a rational discussion only within the unproblematic horizon of a historically concrete life form or an individual way of life" (translation my own).

35. Cf. Habermas, *Justification and Application*, vii: "Going beyond the sterile opposition between abstract universalism and a self-contradictory relativism, I endeavor to defend the primacy of the just (in the deontological sense) over the good. That does not mean, however, that ethical questions in the narrow sense have to be excluded from rational treatment."

interest: reformulating Kant's universalizability principle, he defines the just course of action as the one that is in the interest of everyone affected by it.[36] Given these definitions of ethics and morality, Habermas himself notices that his "discourse ethics" should evidently be called "discourse theory of morality," yet he refrains from renaming it, since the former use has already been established.[37]

Even though for Habermas one must distinguish between the two, morality and ethics need each other, which becomes particularly evident in his *The Future of Human Nature*. Thus he writes, "Under the condition of postmetaphysical thought, the ethical self-understanding of the species, which is inscribed in specific traditions and forms of life, no longer provides the arguments for overruling the claims of a morality presumed to be universally accepted. But this 'priority of the just over the good' must not blind us to the fact that the abstract morality of reason proper to subjects of human rights is itself sustained by a prior ethical self-understanding of the species, which is shared by all moral persons."[38] While morality knows what is just, it has problems mustering the proper motivation on the part of the will to do what is just. It leaves open the question "why be moral?" Here ethics comes in, which inserts the person into the context of a tradition of shared beliefs about the good life: "Moral insights effectively bind the will only when they are embedded in an ethical self-understanding that joins the concern about one's own well-being with the interest in justice. Deontological theories after Kant may be very good at explaining how to ground and apply moral norms; but they still are unable to answer the question of why we should be moral *at all*."[39]

36. Cf. ibid., 32: "If anyone who engages in a corresponding argumentative praxis must make idealizing presuppositions of the sort indicated, then it follows from the normative content of these suppositions of rationality (openness, equal rights, truthfulness and absence of coercion) that, insofar as one's aim is to justify norms, one must accept procedural conditions that implicitly amount to the recognition of a rule of argumentation, (U): 'Every valid norm must satisfy the condition that the consequences and side effects its *general* observance can be anticipated to have for the satisfaction of the interests of *each* could be freely accepted by *all* affected.'"

37. Cf. ibid., vii: "It is my hope that these essays reflect a learning process. This holds at any rate for the explicit distinction between moral and ethical discourses. It is worked out for the first time in the Howison Lecture . . . delivered at Berkeley in 1988 and dedicated to my daughter Judith. Since then it would be more accurate to speak of a 'discourse theory of morality,' but I retain the term 'discourse ethics,' which has become established usage."

38. Habermas, *Future of Human Nature*, 40.

39. Ibid., 4.

However, for Habermas, only morality is universal—consensus on what is just can reasonably be achieved with every rational agent, independent of his or her cultural tradition—while ethics is always restricted to particular communities, the members of which share a common vision of the good, but which cannot necessarily find an agreement on the good with other traditions.

This distinction helps us to understand Habermas' curious claim that he does not give a moral argument against genetic enhancement, and it will also allow us to comprehend his frequent references to an "ethics of the species."[40] The conclusion of his case is not that genetic enhancement is immoral, i.e., unjust, but rather that it compromises the necessary conditions for morality as such, namely the fundamental equality of persons and their ability to attribute their life project and their actions to themselves as their responsible authors. If these conditions are compromised, so is morality. But for Habermas one cannot give a moral argument against a procedure that proposes to abolish morality. Why should we want to understand ourselves as moral beings in the first place? Why should we seek to be just and organize our lives by means of laws and norms? Here, then, when morality as such is at stake, ethics, and with it the question of the good, has to come in again. What do we as members of the particular community comprised of all members of the species consider to be a good life? This is the question posed by Habermas' "species ethics," and it would seem that part of the answer would be that it is good to be moral. It is good to be just. However, to Habermas' mind, philosophy still cannot decide on questions of the good with any cogency. It is ultimately a question left up to our decision: how do we want to understand ourselves? Habermas certainly has a "preference," but at the end of the day he does not think this preference can be universalized. This is why he concludes his booklet with an appeal to how we *feel* and how "we still think today"—both criteria being ultimately contingent:

> Without the emotions roused by moral sentiments like obligation and guilt, reproach and forgiveness, without the liberating effect of moral respect, without the happiness felt through solidarity and without the depressing effect of moral failure, without the "friendliness" of a civilized way of dealing with conflict and opposition, we would feel, or so we still think today, that

40. For instance, the third part of the book's main essay, "The Debate on the Ethical Self-Understanding of the Species," is entitled "The embedding of morality in an ethics of the species" (ibid., 37–44).

> the universe inhabited by men would be unbearable. Life in a
> moral void, in a form of life empty even of cynicism, would not
> be worth living. This judgment simply expresses the "impulse"
> to prefer an existence of human dignity to the coldness of a form
> of life not informed by moral considerations.[41]

If the threat is rightly diagnosed as the abolition of the human person as a moral being, then, in the face of the danger's magnitude, this
conclusion is rather weak. Perhaps his whole argument can be seen as
exemplifying the difficulties connected with the separation between morality and ethics, the just and the good, and maybe even as an implicit
admission that this separation cannot be carried through to the end. If
Habermas admits that justice cannot motivate anyone to be just unless
he also realizes that justice is a good, then we wonder whether his attempts at neatly separating the just from the good as two different realms
does not falter. Or, from the other side, if his moral theory cannot by
itself muster any reason for why we should be moral—for why it is better
to live in a moral world, where people act and interact rather than in a
technocratical world, where people simply behave and function—then
this would certainly speak against his theory.[42]

We are, however, not so much concerned here with the effects of
genetic manipulation on Habermas' moral theory but rather with its
repercussions for the human person. And here it seems to us that Habermas has a genuine insight when he suggests that genetic enhancement 1)
compromises the ability of persons to see themselves as the true authors
of their lives and actions and 2) that it threatens people's fundamental
equality, especially that between the generations.[43] For both of these—re-

41. Ibid., 73.

42. Reflecting on Habermas' thoughts on this issue, Ludwig Siep—it seems rightly—concludes that for Habermas "morality itself is a life form that must and can be
evaluated" (Siep, "Moral und Gattungsethik," 115; translation my own). This idea of
course contains a fundamental flaw, since to evaluate anything we need criteria of
judgment which will have to derive from larger contexts into which we insert that
which we want to judge. Now the moral perspective would seem to be precisely the
broadest perspective possible, inasmuch as by it we look at our life as a whole. The
question "Why be moral?" is thus not an amoral but an immoral question, implying the positive refusal to look at one's life as a whole and to ask how one can live it
well—or, at least, justly.

43. Allen Buchanan's et al.'s remark that genetic engineering may endanger the
unity of the species is not all too far fetched: "We can no longer assume that there will
be a single successor to what has been regarded as human nature. We must consider
the possibility that at some point in the future, different groups of human beings may

lated—claims he draws on the thought of Hans Jonas. We will deal with these then, and their inner cogency and plausibility, believing that if they are true, we will have sufficient ground to call genetic enhancement an immoral procedure that should also be illegal. Hence, we will leave aside for the moment concerns about the transcendental conditions of morality as such and focus on these two concrete claims.

The Distinction between Therapy and Enhancement

The first thing to say is that Habermas' argument presupposes the distinction between therapy and enhancement. A genetic intervention at the embryonic stage that is motivated by a "clinical attitude," aiming at the cure of a disease or malformation of whatever kind, does not pose a problem for him, as one can assume the virtual consent of the person that the embryo "will one day be" (to call the embryo as such a person is too big of a metaphysical commitment for Habermas—but this is not our argument now). Thus, he writes, "In the case of *therapeutic* gene manipulation, we approach the embryo as the second person he will one day be. This clinical attitude draws its legitimizing force from the well-founded counterfactual assumption of a possible consensus reached with another person who is capable of saying yes or no."[44] Individuals, on learning that their parents, by means of a genetic intervention, prevented them from having a heart-disease, will be unlikely to object to their decision. The embryos' future consent can be safely assumed, so that in the therapeutic intervention they are always treated as second persons in an anticipated communicative context. In the case of genetic enhancement, however, this consent cannot be presumed. It is an optional treatment that the

---

follow divergent paths of development through the use of genetic technology. If this occurs, there will be different groups of beings, each with its own 'nature,' related to one another only through a common ancestor (the human race), just as there are now different species of animals who evolved from common ancestors through random mutation and natural selection." Buchanan et al. are quite aware of the difficulties these beings may have still to understand themselves as moral community: "The effectiveness of people's motivation to act consistently on universal moral principles may depend significantly on whether they share a sense of common membership in a single moral community. But whether this sense of moral community could survive such divergence is a momentous question" (Buchanan et al., *From Chance to Choice: Genetics and Justice*, 95). All the more surprising is their eventual endorsement of a liberal eugenics (cf. 378–82).

44. Habermas, *Future of Human Nature*, 43.

persons affected may welcome or not. Who could presume to know that they wanted blue or green eyes, that they wanted black or blond hair, etc.? Insofar as here another person's decision is imposed on them, the embryos are treated from a third person perspective, i.e., as objects that merely undergo the procedure but who are in no position to say yes or no. There is no anticipated communicative context; the embryos are treated as objects of instrumental action rather than as potential subjects in the context of intersubjective communicative action.

Of course, Habermas knows that the distinction between therapy and enhancement is not unproblematic.[45] For instance, it requires us to have some idea of what we mean by "health," which is a concept that is very difficult to define.[46] As the President's Council on Bioethics points out, "most human capacities fall along a continuum, or 'normal distribution' curve, and individuals who find themselves near the lower end of the normal distribution may be considered disadvantaged and therefore unhealthy in comparison with others. But the average may equally regard themselves as disadvantaged with regard to the above average. If one is responding in both cases to perceived disadvantage, on what principle can we call helping someone at the lower end 'therapy' and helping someone who is merely average 'enhancement'?"[47] In other words, there is the problem, for instance, that people come in all shapes and sizes; some people are shorter than others within a range of what is considered

45. Thus, in ibid., 117–18n3, Habermas cites Nicholas Agar who writes, "Liberals doubt that the notion of disease is up for the moral theoretic task the therapeutic/eugenic distinction requires of it" (Nicholas Agar, "Liberal Eugenics"). Agar later published a whole monograph of the same title, in which he also refers to Habermas' argument (cf. Agar, *Liberal Eugenics: In Defense of Human Enhancement*).

46. Cf. the report by the President's Council on Bioethics, *Beyond Therapy*, 15: "Both 'enhancement' and 'therapy' are bound up with, and absolutely dependent on, the inherently complicated idea of health and the always-controversial idea of normality. The differences between healthy and sick . . . are experientially evident to most people. . . . But there are notorious difficulties in trying to define 'healthy' and 'impaired,' 'normal' and 'abnormal.' . . . If one follows the famous World Health Organization definition of health as 'a state of complete physical, mental and social well-being,' almost any intervention aimed at enhancement may be seen as health-promoting."

Evidently, the same problems that there are in defining "health" exist for the correlated concept of "disease," which Nicholas Agar is quick to point out: "The notion of disease turns out to be difficult to define, and without an adequate definition we cannot be certain of what is an appropriate use of gene therapy, and what is not." Nonetheless, for him "it would be a mistake . . . either to give up on defining disease, or to assume that the notion is just too obvious to warrant definition" (Agar, *Liberal Eugenics*, 79).

47. The President's Council on Bioethics, *Beyond Therapy*, 15–16.

normalcy. Below a certain size, however, a person's shortness may be considered a medical condition warranting a therapeutic intervention. Yet, in case he or she is above a specific size, the intervention will be called enhancement. It is clear that to make the distinction between therapy and enhancement in this case, we will have to draw a line at some point. It is also clear that the point where this line will be drawn will depend on the context of a given society, where people in general will be shorter or taller. In other words, it will always be context-dependent and ultimately arbitrary. The same counts for many other possible conditions: when is a heart so weak that an intervention to make it stronger is therapeutic and when do we begin to enhance the strength of a relatively normal heart? The distinction is difficult to make, and therefore Habermas writes that "in any case, *assumed* consensus can only be invoked for the goal of avoiding evils which are unquestionably extreme and likely to be rejected by all."[48] To figure out concretely what these extreme evils are, he considers a task and responsibility of society at large.[49] To him, the "regulative idea" itself, in any case, is very clear and not as such contestable, although it is "surely in need of continuous interpretation": we speak of therapy in cases when we can presume the other's consent; we speak of enhancement when we cannot.[50]

This solution proposed by Habermas is of course itself somewhat circular. We define as therapeutic those interventions for which we can presume the patient's (future) consent. At the same time our presumption of the patient's (future) consent rests on our prior judgment that the intervention is precisely this, namely therapeutic. In doing so we have not advanced a single step. But let us grant that, with a wide field of grey in the middle, there are at least some cases where things are black and white. Thus, Francis Fukuyama, himself a member of the President's Council on Bioethics and thus a co-author of *Beyond Therapy*—a work

48. Habermas, *Future of Human Nature*, 43.

49. Cf. ibid., 43–44: "Thus the moral community which in the profane realm of everyday politics takes on the sober form of democratically constituted nations must eventually believe itself capable of working out, time and again, from the spontaneous proceedings of everyday living, sufficiently convincing criteria for what is to be understood as a sick, or a healthy, bodily existence."

50. Cf. ibid., 90–91: "There is a regulative idea that establishes a standard for determining a boundary, one which is surely in need of continuous interpretation, but which is not basically contestable: All therapeutic genetic interventions, including prenatal ones, must remain dependent on consent that is at least counterfactually attributed to those possibly affected by them."

that emphasizes the difficulties with the distinction between therapy and enhancement—writes at a different place: "While it is the case that certain conditions do not lend themselves to neat distinctions between pathological and normal, it is also true that there is such a thing as health. . . . There is a natural functioning to the whole organism that has been determined by the requirements of the species' evolutionary history, one that is not simply an arbitrary social construction."[51] He brings his case succinctly to the point when he writes, "That the boundary between therapy and enhancement is unclear does not make the distinction meaningless."[52]

Habermas and Jonas: The Question of Domination

Supposing that Fukuyama's conclusion can in principle be granted, what then is the problem peculiar to genetic enhancement as Habermas sees it? How exactly does it compromise the affected persons' ownership of their own life and their equality with others? Here, then, Habermas draws on Jonas and argues that by intervening on the human genome, the biotechnological intervention blurs "the intuitive distinction between the grown and the made, the subjective and the objective—with repercussions reaching as far as the self-reference of the person to her bodily existence." He continues by pointing out that the final result "of this development is characterized by Jonas as follows: 'Technologically mastered nature now again includes man who (up to now) had, in technology, set himself against it as its master.'"[53] Here we recognize Jonas' great concern present throughout *The Imperative of Responsibility,* that we are at a point of losing power over the power we have over nature, the concern, in other words, that our technological power, let loose, turns back on us so as to

51. Fukuyama, *Our Posthuman Future,* 209.

52. Ibid., 210.

53. Habermas, *Future of Human Nature,* 47. Habermas is citing Jonas, *Technik, Medizin und Ethik,* 168. The English translation is by Habermas' translator from the German text. The essay, however, also exists in English, where it is slightly different from the German version: "Now man's impending control over his own evolution is hailed as the final triumph of this power—'nature' now significantly including man himself, reclaiming him as it were from his splendid isolation" (Jonas, *Philosophical Essays,* 145).

The fact that Habermas gets the title of Jonas' book wrong—both in the German and the English version of Habermas' volume it is cited as *Technik, Medizin und Eugenik*—does not as such speak against our thesis that Jonas plays an important role in Habermas' argument here.

dominate us who were initially using it to dominate nature. Habermas shares Jonas' apprehension, particularly when it comes to questions of genetic manipulation of human beings. Here, for Habermas "domination of nature turns into an act of self-empowering of man, thus changing our self-understanding as members of the species." Even though in this particular passage he continues only carefully, proposing that this act may "*perhaps*" touch "upon a necessary condition for an autonomous conduct of life and a universalistic understanding of morality,"[54] one can see later in the book that this is precisely the conclusion of his argument.[55] But what is so grave about this "act of self-empowering of man" for it to have such disastrous effects? At this point Habermas again refers to Jonas who "addresses this concern by asking: 'But whose power is this—and over whom or over what? Obviously the power of those living today over those coming after them, who will be the defenseless objects of prior choices made by the planners of today. The other side of the power of today is the future bondage of the living to the dead.'"[56] By means of genetic programming of the kind for which no later consent can be assumed, one generation exercises an asymmetrical and nonreciprocal power over the other, introducing a fundamental inequality into the relationship between them.

Genetic Determinism?

This, as Habermas points out, is quite independent from the actual effects of the genotype upon the phenotype, i.e., he emphasizes that his argument does not presuppose any genetic determinism, which of course is a faulty theory, simply because many genetic dispositions are not reflected in the phenotype. In addition, the environment has a significant influence

54. Habermas, *Future of Human Nature*, 48.

55. Thus, commenting on his own argument in the post-script, Habermas writes, "This is not itself a moral argument. But it is an argument that appeals to the preconditions for preserving a moral self-understanding of persons as a reason for favoring a species-ethical understanding that cannot be squared with the heedless optimization and self-instrumentalization of prepersonal life" (ibid., 94).

56. Ibid., 48, citing Jonas, *Technik, Medizin und Ethik*, 168. In the English version of Jonas' article, the passage reads: "But of whom is this a power over what and whom? Plainly, of the living over posterity; more correctly, of present men over future men, who are the defenseless objects of antecedent choices by the planners of today. The obverse of *their* power is the later servitude of the living to the dead" (Jonas, *Philosophical Essays*, 145).

on the latter. Habermas shows perfect awareness of these facts,[57] and his argument does not require us to spell out in detail the *exact* influence of the genotype on the phenotype, although the fact that the genotype does have a *certain* influence on the phenotype is presupposed insofar as the procedures of genetic enhancement are chosen precisely because they are deemed efficient. But what matters more than the actual effect on the phenotype is the parents' intention.[58] Upon learning about the prenatal intervention, adolescents will be in danger of seeing their own body as something manufactured, something for which others can and must give an account; they will have difficulties with being "at home" in their body, insofar as alien intentions and expectations are engrained in it. Thus, Habermas writes, "Irrespective of how far genetic programming could actually go in fixing properties, dispositions, and skills, as well as in determining the behavior of the future person, *post factum* knowledge of this circumstance may intervene in the self-relation of the person, the relation to her bodily or mental existence."[59] Following Helmut Plessner, he points out how our bodies are not only something that we have but something that we are.[60] Learning that others have manipulated our bodies without our consent, without entering into a prior communicative

57. Cf. Habermas, *Future of Human Nature*, 53. Having presented Andreas Kuhlmann's observation that there is a difference between parents' general wishful thinking and eugenic programming with which parents impose their ideas on their children, Habermas writes, "To associate this intuition with genetic determinism would be to misconstrue it."

Given Habermas' explicit emphasis on not presupposing genetic determinism, it is surprising how Eduardo Mendieta can claim that Habermas "seems to operate on an unspoken but discernible genetic determinism," without even mentioning the fact that Habermas has raised the objection to himself, attempting also to answer it (Mendieta, "Habermas on Human Cloning," 730).

58. Cf. Habermas, *Future of Human Nature*, 124n54: "It is primarily the intention governing the eugenic intervention that counts. The person concerned knows that the manipulation has been carried out with the sole intention of action on the phenotypic molding of a specific genetic program, and this of course on condition that the technologies required for this goal have proved to be successful."

59. Ibid., 53.

60. Cf. ibid., 50: "A person 'has' or 'possesses' her body only through 'being' this body in proceeding with her life. It is from this phenomenon of being a body and, at the same time, having a body [*Leibsein und Körperhaben*] that Helmut Plessner set out to describe and analyze the 'excentric position' of man. . . . The primary mode of experience, and also the one 'by' which the subjectivity of the human person lives, is that of a body." Cf. Plessner, *Die Stufen des Organischen und der Mensch*.

relationship with us, we may be alienated from our bodies and ultimately from our selves.

Habermas and Jonas: Genetic Engineering vs. Education

Habermas admits that certainly everybody is subject to organic conditions; our body is always a condition of our existence, which has to be accepted and assumed by our freedom. Here, "the situation of the programmed person does not initially differ from that of a person naturally begotten."[61] Problems arise however as soon as "eugenic programming of desirable traits and dispositions . . . commits the person concerned to a specific life-project or, in any case, puts specific restrictions on his freedom to choose a life of his own."[62] Genetic enhancement will be connected with the parents' expectation for the children to lead a certain life, and as long as the adolescents later make these expectations their own, for Habermas there will be no misgivings. However, the possibility of "dissonant cases" cannot be excluded, i.e., the children may later develop interests that are not in accord with their parents' intentions, however noble and benevolent these intentions might have been.[63] In Habermas'

61. Habermas, *Future of Human Nature*, 61.

62. Ibid.

63. Here it may be interesting to have a brief look at Karl Rahner's considerations, which he put down already in the mid-1960s and in which he looks precisely at the basic nature of the parents' intentions. Incidentally, given that for him "man is essentially an inhabitant of 'civilisation' and never merely a factor of nature," and given that thus one may ask whether genetic manipulation is not just another way in which he forms himself culturally, Rahner's ultimately strong criticism of genetic manipulation may be surprising (Rahner, "Problem of Genetic Manipulation," 230). It is similar to Habermas' argument to the extent that it has to make do without a strong concept of human nature. However, it is different and in a sense complementary insofar as it does not look at the issue from the child's but from the parents' perspective. *For them* to be free, they have to be able to relate to their child as the other whom they *accept* and not choose. Particularly the act by which a new human being comes into existence must not be marked by a "neurotic anxiety in the face of fate" (244), which for Rahner is precisely the fundamental motivation that moves parents to genetically manipulate their children: "If man, when confronted with his child, saw only what he had himself planned, he would not be looking at his own nature, nor would he experience his true self which is both free *and* the object of external determination. Genetic manipulation is the embodiment of the fear of oneself, the fear of accepting one's self as the unknown quantity it is. . . . What, in actual fact, is the driving force behind genetic manipulation? What sort of person is driven to it? And the answer would be, in the first place, the *hate* of one's destiny; and secondly, it is the man who, at his innermost level, is in despair

view it is in these cases that one can best see the difference between genetic programming and education. In other words, it is probably natural
for parents to have expectations of their children, but the way in which
these expectations are expressed in either of these cases is fundamentally
different. Habermas points out that "socialization processes proceed only
by communicative action, wielding their formative power in the medium
of propositional attitudes and decisions which, for the adult persons to
whom the child relates, are connected with internal reasons even if, at a
given stage of its cognitive development, the 'space of reasons' is not yet
widely open to the child itself."[64] Given that education takes place within
the realm of reasons, the adolescent can, at least at the point when he
or she comes to the use of reason, accept or reject the process. Education is always interactive and therefore the child will always have "the
role of a second person." For this reason, the "expectations underlying
the parents' efforts at character building are essentially 'contestable.'"[65]
Habermas goes on, "Since even a psychically binding 'delegation' of children can be brought about in the medium of reasons, the adolescents in
principle still have the opportunity to respond to and retroactively break
away from it. They can retrospectively compensate for the asymmetry of
filial dependency by liberating themselves through a critical reappraisal
of the genesis of such restrictive socialization processes."[66] The key point
here is that education, even though it certainly strongly influences the
children and transmits parental expectations, nonetheless in principle
leaves open the later "critical reappraisal" on the part of the adolescents
or young adults, who are able to take a stance *toward* what has happened
to them during the socialization process.

Now Habermas argues that with genetic programming things are
very different. There is evidently no communicative context in which the
child is addressed as a second person, since we are not speaking about
interventions for which a consent could reasonably be presumed. And
what is more, from the point of view of the child, a "critical reappraisal"
will not be possible. The instrumental determination "does not permit
the adolescent looking back on the prenatal intervention to engage in
a *revisionary* learning process. *Being at odds with* the genetically fixed

---

because he cannot *dispose* of existence" (245).

64. Habermas, *Future of Human Nature*, 61–62.

65. Ibid., 62.

66. Ibid.

intention of a third person is hopeless. The genetic program is a mute and, in a sense, unanswerable fact."[67] In this case the attempt on the part of the persons who have been manipulated to break away from the other's intention for them is bound to fail insofar as this intention has become incarnate in their very bodies, i.e., in their very selves.[68] The result is that "someone who is at odds with genetically fixed intentions is barred from developing, in the course of a reflectively appropriated and deliberately continued life history, an attitude toward her talents (and handicaps) which implies a revised self-understanding and allows for a *productive* response to the initial situation."[69] Genetic enhancement is an intervention on the body of the child at the very moment of the body's constitution, thus determining the person's bodily identity. Now the body is not only something the person *has* but also something that the person *is*, and hence the genetic intervention touches his or her very being and identity. Education, in contrast, assumes the child's initial bodily makeup among the given factors of the person's identity and proceeds from there. In the context of education, persons can always take a reflexive stance toward their educators: they can say, "I am here and you are there." They can say, "No." They have the option of changing their stance over time, to accept or reject what has been handed on to them. Genetically modified persons have no such option with respect to the intervention, which had an effect on their very identity, so that a reflexive stance cannot be taken. The modification cannot be rejected but is a necessary part of who these persons are. Now this modification was instrumentally imposed on them outside any communicative context, so that the relation between the parents and the children will take on the character of designers to products. In such a relation the children are instrumentalized and any relationship of equality is precluded. The children owe their very identity, their *Sosein*, their *being-this-way*, to the deliberate choice of other people. By giving them their very *being-this-way*, by determining their identity, these others are no longer their equals. They *owe* to them some of their defining traits, which were imposed on them without any virtual

---

67. Ibid.

68. Cf. the very useful reflections by Bernard G. Prusak on Habermas' argument: "If a person experiences her body as manufactured or produced, she may resent her parents who not only had her but *made* her; but she may also resent her body itself as incarnating alien intentions" (Prusak, "Rethinking 'Liberal Eugenics,'" 36).

69. Habermas, *Future of Human Nature*, 62.

or actual communicative relationship, so that there is a relationship of indebtedness that precludes equality.[70]

Besides impeding the equality between parents and children, genetic modification for Habermas is likely to have another deleterious effect. By knowing of their indebtedness for their identity, according to Habermas, genetically modified persons will find it difficult to regard themselves as the responsible authors of their lives, since other human beings have made nonnegotiable, fundamental decisions for them, decisions which they are unable to challenge as these have become part of their very selves and for which they can *blame* others. Here the distinction between chance and choice, between what simply happens to us as the result of chance or fate and what is done to us by the deliberate intention of other persons, becomes central. In this context Habermas cites Ronald Dworkin as saying, "that crucial boundary between chance and choice is the spine of our ethics and our morality, and any serious shift in that boundary is seriously dislocating."[71] Commenting on Dworkin, Habermas writes, "Shifting the 'line between chance and choice' affects the self-understanding of persons who act on moral grounds and are concerned about their life *as a whole*."[72] For Habermas, this line has a determinate influence on how we understand ourselves as moral beings and as members of the species, which in turn decides on "whether or not we may see ourselves as the responsible authors of our own life history and recognize one another as persons of 'equal birth,' that is, of equal dignity."[73] To make explicit why this line between chance and choice is so important and why shifting it can have serious consequences for our capacity to regard ourselves as the responsible authors of our lives, we can read on in Dworkin's book. A few paragraphs further, though no longer cited by Habermas—for whom these considerations are probably self-evident—Dworkin expresses very clearly the difficulties Habermas seems to have in mind:

> We now accept the condition in which we were born as a pa-
> rameter of our responsibility—we must make the best of it that

70. The truth of C. S. Lewis' insightful affirmation, namely that "what we call Man's power over Nature turns out to be a power exercised by some men over other men with Nature as its instrument," would thus find its preeminent expression precisely in the procedure of genetic enhancement (Lewis, *The Abolition of Man*, 69).

71. Dworkin, *Sovereign Virtue*, 444; cf. Habermas, *Future of Human Nature*, 28, where the quote is directly translated from the German.

72. Habermas, *Future of Human Nature*, 28.

73. Ibid., 29.

we can—but not as itself a potential arena of blame, except in those special cases, themselves of relatively recent discovery, in which someone's behavior altered his embryonic development, through smoking, for example, or drugs. Otherwise, though we may curse fate for how we are, as Richard Crookback did, we may blame no one else. . . . How would all this change if everyone was as he is through the decisions of others, including the decision of some parents not to intervene but to let nature take its course?[74]

Dworkin, to be sure, is not arguing here *against* genetic manipulation. He is only asking the question whether we can feel completely responsible for our lives if for very important facts of our identity we can actually blame other people, not just chance, fate, or Providence. Habermas' answer to this question is a "no,"[75] and in what follows we will explore in a still deeper fashion how he argues for his position.

As we have said, in the effort to make his case, it is crucial for Habermas to distinguish genetic programming from education. For our purposes, it is significant that here again he acknowledges his dependence on the thought of Hans Jonas, even though this time he does so more indirectly. Thus, he writes regarding a person who has been genetically manipulated that his or her "situation, by the way, is not unlike that of a clone who, by being modeled on the person and the life history of a 'twin' chronologically out of phase, is deprived of an unobstructed future of his own."[76] In a note to this latter affirmation Habermas references the same article by Hans Jonas from which he has cited before and which treats

---

74. Dworkin, *Sovereign Virtue*, 445–46.

75. Cf. Habermas, *Future of Human Nature*, 63: "Eugenic interventions aiming at enhancement reduce ethical freedom insofar as they tie down the person concerned to rejected, but irreversible intentions of third parties, barring him from the spontaneous self-perception of being the undivided author of his own life."

Cf. the succinct way in which Erik Malmqvist summarizes Habermas' argument: "In order to be autonomous, he argues, one has to be guided by intentions and aspirations that are truly one's own. This 'capacity of being oneself' in turn presupposes one experiences one's own body in a certain way, namely as originating from a natural process—as something whose beginning lies beyond what is at human disposal. Habermas suggests that this experience of being at home in one's body might elude the person who has been prenatally engineered by her parents. Her body might no longer appear to her as her own, as the continuation of natural fate, but as the vehicle of someone else's intentions. Consequently, she might not regard herself as 'the undivided author of her own life'" (Malmqvist, "Back to the Future: Habermas's *The Future of Human Nature*," 5).

76. Habermas, *Future of Human Nature*, 62–63.

of both cloning and genetic engineering. One of the central points of this article is the very problem of nonnegotiable expectations—which is also the main issue of Habermas' argument. Thus, in the passage referred to (but this time not cited directly), Jonas warns that a clone would be impaired in the "spontaneity of becoming oneself," because "the known donor archetype . . . will dictate all expectations, predictions, hopes and fears, goal settings, comparisons, standards of success and failure, of fulfillment and disappointment, for all 'in the know'—clone and witness alike; and this putative knowledge must stifle in the pre-charted subject all immediacy of the groping quest and eventual finding 'himself' with which a toiling life surprises for good and for ill."[77]

In that same passage Jonas, just like Habermas, points out that what matters is not the actual effect of the genotype on the phenotype but the *intention* with which the person is made.[78] In the case of clones, they are modeled on another person, their older "twin," who for them will necessarily be the idea and standard, an idea which must be "tyrannical" at that: "Note that it does not matter one jot whether the genotype is really, by its own force, a person's fate: it is *made* his fate by the very assumptions in cloning him, which by their imposition on all concerned become a force themselves. It does not matter whether replication of genotype really entails repetition of life performance: the donor has been chosen with some such idea, and that idea is tyrannical in effect."[79] For Jonas, clones have been robbed of their freedom, because they see what others expect to be their future walking in front of them in the person of their older twin. Others have decided for them "the question 'who am I?', which must be a secret to the seeker after an answer and can find its answer only with the secret there as condition of the search—indeed a condition of *becoming* what may *then be* the answer. The spurious manifestness at the beginning destroys that condition of all authentic growth. . . . In brief, he is antecedently robbed of the *freedom* which only under the protection of ignorance can thrive."[80] The parallel between the clone and the person genetically modified is clear: independent of the effects of the genotype on the phenotype, they are, in their bodies,

77. Jonas, *Philosophical Essays*, 161.

78. Thus Habermas writes, "The only thing that counts for the psychical resonance of the person concerned is the intention associated with the programming enterprise" (Habermas, *Future of Human Nature*, 63).

79. Jonas, *Philosophical Essays*, 161.

80. Ibid., 162.

confronted with the expectations of others for their lives.[81] For clones, of course, this is still much more extreme, since they see those expectations in flesh and blood in the person from whom they were copied, while for genetically programmed persons, there is simply the knowledge that others have intervened on their body so that they would suit their purposes, thus—in Habermas' words—reducing their "ethical freedom insofar as they tie down the person concerned to rejected, but irreversible intentions of third parties, barring him from the spontaneous self-perception of being the undivided author of his own life."[82] In other words—to put it positively—the contingency of the beginning, that is, the moment of the body's constitution, is a crucial condition for the ownership that persons have of their lives, for their spontaneity and freedom.

It is interesting to note that when reading along in Jonas' article cited by Habermas, an article which is not exclusively about cloning but also about genetic engineering, we already find the core idea that for Habermas distinguishes genetic intervention from the socialization process; the latter allows for a later critical reappraisal, while the former does not. This is exactly the same point that Jonas already made in his 1974 article referred to here. Thus, he is critical of the planned perfectibility of positive eugenics insofar as our planning is always shortsighted, lacking the proper criteria and wisdom for our decisions together with the necessary "forecast of how the selection will work out in the continued combinations of reproduction."[83] He concludes that "except for the most unequivocal objects of *negative* eugenics, surely in the dreamland of positive genetic perfectibility, we are *not* buying greater certainty with surrendering the unplanned for the planned."[84] In a footnote to this latter statement, Jonas then points out that in fact there is one realm in which planning for our descendants has its proper space: education. "The place

81. On the question of the relation between genotype and phenotype and the importance of the parental intentions, also see Kass, *Toward a More Natural Science*, 68, who in this case discusses the situation of a clone: "But is the cloned individual's future really determinable or determined? After all, only his genotype has been determined; it is true that his environment will exert considerable influence on who and what he becomes. Yet isn't it likely that the 'parents' will seek to manipulate and control the environment as well, in an attempt to reproduce the person who was copied? For example, if a couple decided to clone a Rubinstein, is there any doubt that early in life young Artur would be deposited at the piano and encouraged to play?"

82. Habermas, *Future of Human Nature*, 63.

83. Jonas, *Philosophical Essays*, 153.

84. Ibid.

for planning with regard to perfectibility, and thus for the shortsight-edness clinging to all planning, is education. There we do impose our necessarily shortsighted image on the forming individual and commit blunders together with our good according to our lights of the hour: but there, in conditioning, we also impart, at least do not foreclose, the means for revision, and certainly leave the inherited nature what it is."[85] Thus Jonas grants that education is in fact "conditioning," but at the same time it provides the means for revision, or, in Habermas' words, it allows for a later "critical reappraisal" on the part of the adolescent or young adult.

In this sense, Nick Bostrom's criticism of Habermas and Jonas does not touch the core of their case. Bostrom writes, "Jürgen Habermas, in a recent work, echoes Jonas' concern and worries that even the mere *knowledge* of having been intentionally made by another could have ru-inous consequences. . . . A transhumanist could reply that it would be a mistake for an individual to believe that she has no choice over her own life just because some (or all) of her genes were selected by her parents. She would, in fact, have as much choice as if her genetic constitution had been selected by chance. It could even be that she would enjoy signifi-cantly *more* choice and autonomy in her life, if the modifications were such as to expand her basic capability set."[86] Neither Habermas nor Jonas claim that genetically engineered human beings would have *no* choice over their life at all. What they affirm is rather that the knowledge of owing part of one's very identity to the arbitrary decisions of peers would be a liability in the task of making one's life completely one's own be-cause one could always blame or would always have to credit others for characteristic marks that form part of who one is without ever being able to make these characteristics really one's own. Maureen Junker-Kenny summarizes Habermas' position succinctly when she writes, "The sense of responsibility for one's own life is compromised by being able to blame the intrusion of another into the core of one's physical being, and conse-quently, of one's self-understanding."[87]

85. Ibid., 153n12.

86. Bostrom "In Defense of Posthuman Dignity," 211–12.

87. Junker-Kenny, "Genetic Enhancement as Care," 6. This is also the response one can give to Nicholas Agar's objection against Habermas: "Habermas's sharp distinction between environmental and genetic influences does not match the reality of human development, however. There are many environmental improvements that do not grant any realistic right of reply" (Agar, *Liberal Eugenics*, 116–17). Many decisions that parents make for their young children are in fact irreversible in their consequences and hence "do not grant any realistic right of reply." And yet, in these cases one can

Habermas proposes that to be free to reassess what others do to us in education, we need a point that is withdrawn from the intervention of peers. If individuals had to see themselves entirely as the product of "a suffered socialization fate," that is, as entirely the result of what others have done to them, then they would have to see their "'self' slip away in the stream of constellations, relations, and relevancies imposed upon the formation process."[88] Now it is precisely our naturally-grown bodily existence born from a contingent beginning that provides us with such a point of retreat—an original "what-we-are"—from which we can take a stance toward what happens to us. As Habermas puts it, "We can achieve continuity in the vicissitudes of a life history only because we may refer, for establishing the difference between what *we* are and what happens *to us,* to a bodily existence which is itself the continuation of a natural fate going back beyond the socialization process. The fact that this natural fate, this past before our past, so to speak, is not at our human disposal seems to be essential for our awareness of freedom."[89] We are reminded here of Archimedes' "Give me where to stand and I shall move the earth." To move the earth we need a place outside the earth. We can only be free with respect to anything, free to move it, if we can in some ways stand outside of it.[90] It is Habermas' provocative suggestion that we can be free in the realm of human affairs only if we have a place to stand that is withdrawn from direct human intervention. Such a place is provided by our contingent beginning, expressed in our untampered bodily existence. When we genetically manipulate individuals or use other, already available eugenic methods, it is as if we were to remove the sure ground from under their feet, the ground on which they can stand to actualize their freedom.

---

either safely assume the children's consent or at least the intervention stays outside of them, so that there remains one point at the core of their identity—the moment of the constitution of their physical genetic makeup—that is withdrawn from the intrusion of peers.

88. Habermas, *Future of Human Nature,* 60.

89. Ibid.

90. Cf. Hannah Arendt who traces back the extraordinary success of natural science with its incredible increase of power over nature to the discovery of the "Archimedean point," a fundamental change of perspective by means of which the whole earth can be looked at from the outside, namely from the viewpoint of the universe (Arendt, *Human Condition,* 257–68). Similarly we may say that individuals are free "to move the earth"—to be the responsible authors of their lives—only if they can stand on a point that is untouched and removed from the intrusion of others.

Dependence as Part of the Human Condition

Although we are always conditioned beings who have not made them-selves, for Habermas it makes an important difference whether our being this or that way is the result of chance or God on the one hand or of another human being on the other.[91] As already Hans Jonas noted, the designer who decides about our traits, who "manufactures" us, domi-nates us in a way that is incompatible with human equality. If all are of "equal birth," owed to chance or to God, all are equal, while if some are designed by others, they are brought into a previously unheard of posi-tion of dependence. Now of course we are all dependent, and no one has made himself. Robert Song rightly stresses this fact when, unconvinced by Habermas' emphasis on the need for our sole authorship of our own life history, he writes, "Co-authorship is an inescapable feature of our lives. We find ourselves in a situation in which not only the random mix-ing of our parents' genes but also the conscious decisions they took form the context of our lives . . . . Not only are we not the sole authors of our life history, that is not even an aspiration to aim for."[92] We may grant to Song that Habermas, when speaking of the danger that genetically

91. This is the concern Habermas voices at the end of his discourse on "Faith and Knowledge": "Because he is both in one, God the Creator and God the Redeemer, this creator does not need, in his actions, to abide by the laws of nature like a technician, or by the rules of a code like a biologist or computer scientist. From the very beginning, the voice of God calling into life communicates within a morally sensitive universe. Therefore God may 'determine' man in the sense of enabling and, at the same time, obliging him to be free. Now, one need not believe in theological premises in order to understand what follows from this, namely, that an entirely different kind of de-pendence, perceived as a causal one, becomes involved if the difference assumed as inherent in the concept of creation were to disappear, and the place of God be taken by a peer—if, that is, a human being would intervene, according to his own preferences and without being justified in assuming, at least counterfactually, a consent of the con-cerned other, in the random combination of the parents' sets of chromosomes. This reading leads to the question I have dealt with elsewhere: Would not the first human being to determine, *at his own discretion*, the natural essence of another human being at the same time destroy the equal freedoms that exist among persons of equal birth in order to ensure their difference?" (Habermas, *Future of Human Nature*, 115).

Although Buchanan et al. approve of liberal eugenics, they nonetheless propose cautionary considerations akin to those advanced by Habermas: "Even if an individual is no more locked in by the effects of a parental choice than he or she would have been by unmodified nature, most of us might feel differently about accepting the results of a natural lottery versus the imposed values of our parents" (Buchanan et al., *From Chance to Choice*, 177–78).

92. Song, "Knowing There Is No God," 201.

modified persons "might no longer regard themselves as the sole authors of their own life history,"[93] is using a possibly misleading expression. We are never the sole authors of our own life. But from his whole argument it is clear that Habermas does not want to propose the "ideal" of a monadic and isolated existence. Rather, he means to assert that in order for us to regard ourselves as *responsible* authors, there needs to be some point withdrawn from the direct intervention of peers.

Habermas admits that children are always dependent on their parents for their lives, but this, he argues is fundamentally different from owing certain traits to one's parents' willful decision. For the latter, one can ask an account, while it would seem "abstract" to hold parents responsible for one's existence.[94] In case of natural conception and development, even though we receive our genetic makeup from our parents, what ultimately comes of it is contingent: we do not owe our *being-this-way* to any deliberate act of our parents but rather to chance or to God, but in either case not to any other human being who would have to give us an account for it and of whom we could justly ask such an account. In the case of genetically enhanced children, these could ask their parents why they had such a long nose, why they had blue eyes and not brown ones and why they were smart but not still smarter. Their parents would have to be answerable to these questions, which are, however, ultimately unanswerable.

Excursus: The Question of Artificial Procreation

An important question we must ask here is whether what Habermas says about persons' *being-this-way* does not also apply in some ways to their *being* plain and simple. Is it only when parents predispose some of their children's traits that they introduce an unacceptable aspect of inequality into the relationship to their children, or is this not also already the case when they will their existence in too strong and direct a manner—namely by means of artificial procreation?[95] Along this line

---

93. Habermas, *Future of Human Nature*, 79.

94. Cf. ibid., 64.

95. Cf. the concern voiced by O'Donovan, *Begotten or Made?* 2: "A being who is the 'maker' of any other being is alienated from that which he has made, transcending it by his will and acting as the law of its being. To speak of 'begetting' is to speak of quite another possibility than this: the possibility that one may form another being who will share one's own nature, and with whom one will enjoy a fellowship based on

of thought, many years before Habermas voicing himself on this issue, other philosophers like Robert Spaemann and Martin Rhonheimer, basing themselves on the Vatican's Instruction *Donum vitae,* have reflected on the implications of artificial procreation and have seen problems very similar to the ones that Habermas detects with genetic enhancement.[96] They convincingly argue that there is a problem not only with parents deciding on the genetic makeup of their children, on their *being-this-way,* but already with their direct will for their children simply to *be* at all costs, an emphatic will that finds expression in practices such as in-vitro fertilization. In natural procreation children are not the product of an act of manufacture. Rather, they come into being while their parents are doing something else. Ideally, they are the *by-product* of an act of love. Whether their parents were thinking of them or not while they were together, the children are not their direct product: they are "begotten not made,"[97] and hence their very existence, not only their *being-this-way,* has the character of something contingent for which no other human person owes any account. The same cannot be said of children whose existence comes about artificially. Here the only reason for the action that is performed in bringing them about is precisely to produce them. For this action of artificial procreation their parents owe them a justification. They can meaningfully ask them why they wanted them to be. But the existence of a human person is nothing for which other human beings can give any justification. It is beyond them. By turning the parent-child relationship into a relationship between producer and product already artificial procreation instrumentalizes the child in the same way in which genetic engineering does for Habermas. In a little excursus we will look at this issue in more detail.

While it is certainly a good thing for spouses to desire a child, which constitutes the crowning fulfillment of their loving union, it can be argued that the production of the child *in-vitro* introduces a disorder into the relationship between child and parents, much the same as the one that Habermas claims is introduced by the parents' manipulation of their children's traits. As the instruction *Donum vitae* puts it, the technological

radical equality."

96. Cf. Congregation for the Doctrine of the Faith, *Donum vitae*; Spaemann, "Kommentar"; Rhonheimer, "The Instrumentalization of Human Life" (originally published in German in 1992).

97. Cf. Spaemann, "Gezeugt, nicht gemacht." For a theological perspective, see: O'Donovan, *Begotten or Made?* and Marco Hofheinz, *Gezeugt, nicht gemacht.*

dominion of a human person's origin results in an unjust inequality between the parents and their child: "Homologous IVF and ET is brought about outside the bodies of the couple through actions of third parties whose competence and technical activity determine the success of the procedure. Such fertilization entrusts the life and identity of the embryo into the power of doctors and biologists and establishes the domination of technology over the origin and destiny of the human person. Such a relationship of domination is in itself contrary to the dignity and equality that must be common to parents and children."[98] But how exactly does this element of inequality enter the relationship between parents and children? Why is this technological domination contrary to "the dignity and equality that must be common to parents and children"? To see how artificial procreation is an act of injustice toward the children thus conceived, we will examine the parent-child relationship and argue that the question of whether children are conceived naturally or artificially makes a big difference for the very nature of this relationship.

To understand the difference in the parent-child relationship in the two cases, it is helpful to look at the difference in the way the child "results" from natural intercourse versus artificial procreation. We have reason to say that the child who is conceived during the conjugal act is the *fruit* of an act of love and comes to the parents as a *gift*. In the words of Gabriel Marcel, he or she "no more belongs to us than we do to ourselves."[99] Put differently, the child is like "a guest who comes from afar," whom one can welcome properly only when recognizing the "unexpected and gratuitous initiative of Another."[100] The parents may in fact hope to conceive a child during the conjugal act, but, at least so long as the spouses choose to perform this act with the right attitude, it is always more than an act of human generation.[101] It is an act of spousal love that spouses would perform and could rightly perform also if they knew that

---

98. Congregation for the Doctrine of the Faith, *Donum vitae*, II, B, 5.

99. Cf. Marcel, *Homo viator*, 120. In the same chapter Marcel warns us that fatherhood "degenerates as soon as it is subordinated to definitely specified purposes, such as the satisfaction of ambition through the medium of the child treated as a mere means to an end" (116).

100. Melina, *Building a Culture of the Family*, 168.

101. Cf. Rhonheimer, "Instrumentalization," 162–63: "What the spouses voluntarily do when they engage in intercourse (with or without an explicit desire for a child) can be described as a reciprocal self-gift, specifically in the totality of their being man and woman. . . . 'The generation of a child' is not therefore in every case an adequate description of what the spouses *do* when they have intercourse."

they were sterile.[102] All that is required is that they do not hinder it from being fruitful, i.e., they need an intentional openness to new life, but they do not always need directly to intend to procreate new life. In fact, if the act were to serve solely as means to procreation and not also the expression of spousal love, it would itself be disordered.[103] It remains meaningful even if there is no chance of pregnancy. As Rhonheimer concludes, "The conjugal act . . . is not truly a 'means' for reaching the goal of 'a child.'"[104]

Children conceived by the loving union of their parents can thus consider themselves the *fruit* of their parents' love, and not as their *product*. These children can say that they are because their parents loved each other.[105] Should they ever find themselves in an existential crisis, loathing their own lives, and ask their parents why they "made" them, these would not be obliged to give any account. The children came to be "while they were doing something else."[106] And yes, maybe the parents were delighted at their coming to be, and maybe they were hoping for their coming to be, but what they were doing when they came to be was not "producing another human being," but rather performing an act of spousal love. These parents do not owe their children any existential account for their being, nor do the children owe their being to the causal volition of their parents. They are the fruit of their parents' love, not the product of their making.

Things are different with artificial procreation. What is directly and exclusively intended in the act of artificial procreation is precisely

102. Cf. ibid., 162: "A couple . . . that performs the conjugal act with the desire to have a child, even in the case where a lack of procreative results is certain, would nevertheless continue to perform such acts (in any case they would not abstain because of their infertility), because these acts of sexual union in no way lose their sense or purpose due to their lack of procreative results."

103. Cf. ibid., 164: "The more, however, the conjugal act is carried out solely with the *intention to generate a child* . . . , the more the natural procreative act becomes functionalized to the end of fulfilling this desire for children, which is to say the more it resembles IVF."

104. Ibid., 162.

105. Cf. ibid., 167: "A child generated naturally . . . could never say: 'I exist *because* you wanted me.' In fact, in the generation of this child there was no causative will, only the *desire* for a child. . . . He could say: 'I am, and only for this reason, that you love one another and have given each other the testimony of this love.'"

106. Cf. Spaemann, "Kommentar," 93: "If one day a child were to ask of his parents an account for his perhaps unhappy existence, the parents would not need to give this account. They did not 'make' the child. He came into being naturally, while they were doing something else" (translation my own).

the coming to be of a new human being. Here, the children could not possibly be thought of as the *fruit* of some other activity but they are precisely the exclusively intended *product* of the activity, i.e., the procedures involved in artificial procreation. If attempts of in-vitro fertilization and embryonic transfer repeatedly failed to achieve pregnancy, the woman would eventually stop submitting herself to this procedure. The procedure makes sense only insofar as it may lead to a pregnancy. Quite evidently, it is solely a means to this end, because, if it fails to achieve a pregnancy, the woman will abandon it.[107] A child conceived by means of this procedure is the product of a technological intervention, not the fruit of the mutual self-giving of the spouses. This is precisely the reason why *Donum vitae* calls this procedure illicit: "In reality, the origin of a human person is the result of an act of giving. The one conceived must be the fruit of his parents' love. He cannot be desired or conceived as the product of an intervention of medical or biological techniques; that would be equivalent to reducing him to an object of scientific technology."[108]

Now if children born after artificial procreation were to ask their parents: "Why am I," the response could not be, "Because we loved each other," but only, "Because we wanted you." The children's existence, as the product of a technological intervention, comes to be dependent on their parents' will in an existential way.[109] "The 'goodness' of the new human life is made dependent, *in the act of the decision for IVF and in the acts that effect the procedure,* on its 'being desired.'"[110] The children born of artificial procreation are only because their parents wanted them. Had they not wanted them, they would not have done what they did. This existential dependence on the will of other human beings, however, violates the fundamental equality of human beings. Such a dependence is

107. Cf. Rhonheimer, "Instrumentalization," 162: "A couple that as a last resort decides for IVF, carries out a series of actions whose purpose is manifestly to bring to fulfilment their desire for a child. If this desire didn't exist, or if after several attempts the couple concluded that reproductive medical intervention would not be successful, they would no longer carry out the actions involved in IVF (given that these actions, when the goal is seen as unreachable, would in effect lose their sense or purpose)."

108. Congregation for the Doctrine of the Faith, *Donum vitae*, II, 4 c.

109. Cf. Rhonheimer, "Instrumentalization," 166–67: "A child generated in this way . . . in relation to his parents, could say: 'I exist because you *wanted* me, and only because of this.' In effect, the existence of this child would depend on the will of his parents—it is a function of this will."

110. Ibid., 156.

bearable only in front of God, but not in front of other human beings.[111] When I see my life before God, I can say, "I am because I am loved; I am because God wanted and desired me to be." This is not only bearable but even reassuring. However, there is no equality between God and the human person and there was never meant to be any. Besides, we know that God does not make mistakes and that he does not change his mind. But before other people we do not want to say, "I am because you desired me." What if they were to cease desiring me? What if they were to say they made a mistake? What if I do not live up to their expectations?

We all want to be loved and desired for who we are, for the very fact that we are and not simply because someone else wanted us.[112] We want to feel loved and desired for the very fact that we are and not have our being depend on the fact that we have been desired. If children, conceived in-vitro, one day were to ask their parents for the reasons of their perhaps miserable existence, their parents would in fact owe them an account, because they came to be not only by the simple wish of their parents but by their will to have this wish come true at any cost. For this will, they have to render them an account. But human beings cannot give an account for the existence of other human beings.[113] The very idea of human dignity means that human life is something to be respected unconditionally. We cannot give criteria for the justification of human life. Human life is precious because it is. If we had to give criteria for its justification—for instance, the parents' desire—we would make it conditional on

111. Cf. ibid., 168: "A child *wanted by his parents* and generated in vitro, . . . would have a measure of existential dependence with respect to his parents that contradicts both his fundamental *equality* as a person and his *freedom*. Only before God is such a dependence tenable, becoming, indeed, the foundation of one's freedom."

112. Cf. ibid., 168: "Procreation according to nature recognizes the gift character of a new life . . . as well as the fact that human life has no need to be justified by being desired. . . . The opposite attitude would go against the most fundamental principle of *justice*, the 'Golden Rule' ('that which you would not want done to you, don't do to others!"). It would therefore threaten the identity of the agent himself, given that every person wants to be recognized by others, not because his existence corresponds to the desire or pleasure of others. . . . Instead, he demands this acceptance or recognition on the pure and simple basis of his existence."

113. Cf. Spaemann, "Kommentar," 93: "Things are different with the test-tube baby. He has been forced into life. He has been 'made.' *He is a product not just of his parents' wish but of their will to fulfill this wish cost what it may.* For this will, the parents owe the child justification. But how do human beings want to justify the existence of other human beings? This is something that essentially exceeds everything for which we possess criteria of justification. Therefore, children must not be degraded to become our products" (translation my own).

these criteria and hence violate its dignity, since having dignity precisely means being worthy of *unconditional* respect and recognition.[114] A being that is worthy of unconditional respect must not be forced into being, as the product of a technician's making, conditional on the parents' desire. No, such a being can only be received as a gift, as the fruit of the parents' mutual self-giving and conditional only on God's own will to call him or her into being: "In his unique and unrepeatable origin, the child must be respected and recognized as equal in personal dignity to those who give him life. The human person must be accepted in his parents' act of union and love; the generation of a child must therefore be the fruit of that mutual giving which is realized in the conjugal act wherein the spouses cooperate as servants and not as masters in the work of the Creator who is Love."[115]

Habermas' Response to Objections

But let us now return once more to Jürgen Habermas himself. In the postscript to his work added to its later editions, he responds to a number of objections that have been leveled against it after its first publication. In the hope that they may help us to understand better what he wants to say, we will briefly deal with four of these, which we consider to carry some particular weight.[116]

114. Cf. Rhonheimer, "Instrumentalization," 158–59: "One would then consider the satisfaction of such a desire [for a child] to be 'good' not because a new life has begun and the parents have had a child, but precisely because now the desire to have a child has been fulfilled. A 'wanted child' in *this* sense would amount to a degradation of human life, because it implies an acceptance or recognition by the parents of this life that is *conditioned*. What makes being a 'couple with children' to be a better state must not be the *fulfillment of a desire*, but the simple, and perhaps even unexpected, *existence of the child*."

115. Congregation for the Doctrine of the Faith, *Donum vitae*, II, 4 c.

116. Among the secondary literature referring to Habermas' *Future of Human Nature* there is no lack of texts that charge Habermas with a blunt naturalism, a criticism that is of course beside the point. Thus Fredrik Svenaeus writes, "The feeling of awe in the presence of nature is, in the end, what brings Habermas to resist embryonic stem-cell research in all its forms" (Svenaeus, "Heideggerian Defense of Therapeutic Cloning," 53). For an argument based on an almost "classical"—but of course unten-able—person-nature dualism, see Bayertz, "Human Nature: How Normative Might It Be?" 131–50. Also see Fenton, "Liberal Eugenics and Human Nature," 40, who under-stands herself to be criticizing Habermas when she writes, "The implication drawn in the arguments against liberal eugenics is that what is manufactured is morally inferior to what is natural, that natural processes in certain areas of human life ought to be

There is first and foremost the important objection that Habermas' argument rests on the possibility of dissonant intentions on the part of the child, i.e., that it applies only to cases in which future consent cannot be presumed. However, it seems that one can presume consent not only when it is a matter of treating an illness or malfunction, i.e., not only in the case of negative eugenics, but also in some cases of positive eugenics, namely when what is at stake is not a *particular* quality, like a particular talent, which may indeed limit the adolescent's choices, but a *general* quality, like physical strength or intelligence, that only serves to open up a greater range of options to the child.[117] In a what seems to us thoughtful response, Habermas notes that with our finite intellects we are inherently unable to predict which capacities will be to the true future benefit of our children, and which ones may actually turn out to be a curse. Not even general qualities may turn out to be an unconditional good in the other person's unpredictable and open life-story. He gives the example of a good memory and explains that at times it can actually be good for someone if he or she is able to forget. Similar things can be said about intelligence. Superior intelligence is not always an unambiguous good. It could have bad effects on the person's character or draw the envy of peers and thus be a source of suffering.[118] The human condition is ambiguous enough that no inheritable or genetic disposition or capacity whatsoever could always be called an unconditional good. As Habermas puts it, "Even in the best of cases, our finite spirit doesn't possess the kind of prognostic knowledge that enables us to judge the consequences of genetic interventions within the context of a future life history of another human being."[119] Hence, even with the enhancement of general qualities we cannot presume the adolescent's consent.

Another objection probes and challenges again the distinction between genetic manipulation and education. In educating their children,

---

protected from the encroachment of unnatural human science. But there is nothing intrinsic to the natural/unnatural distinction to warrant this claim. The natural is not intrinsically good, the unnatural not intrinsically bad."

117. Here Habermas in particular refers to Birnbacher, "Habermas' ehrgeiziges Beweisziel," 121–26; cf. Habermas, *Future of Human Nature*, 85.

An objection along these lines is also raised by Viano, "Antiche ragioni per nuove paure," 292: "What would a son of parents say who, although they could have done it, did not want to give him a good memory, which he then could have used however he wanted?" (translation my own).

118. Cf. Habermas, *Future of Human Nature*, 85–86.

119. Ibid., 90.

parents often seem to make choices for their children without being able to assume their consent, especially when the children are still very young and do not yet have the use of reason. For instance, a musical talent, unless it is favored early on, will go unused. Thus parents may train their children at a young age to further this or that talent, even prior to their being able to give their consent or formulate their own wishes. Genetic enhancement, one could say, is simply a form of education by other means, as both aim at making sure that our children turn out well and both include irreversible choices from our part for which we cannot assume our children's consent.[120] Habermas tries to take on this challenge by noticing that certainly education can go wrong. Instead of taking place in a communicative setting where it introduces our children to reality as a whole and aims at making them autonomous and responsible, it could take place in an instrumental way, where parents try to shape their children according to their prefixed idea. As Maureen Junker-Kenny explains, one of the main attitudes proper to a healthy understanding of education is captured in the biblical commandment "Thou shalt not make an image of me."[121] We must let go of the image we have of our children and help them discover who they are. The fact that at times education becomes the way for parents to force their own ideas on their children can hardly be used as a justification of genetic enhancement. As Habermas formulates it very succinctly, "Insofar as they can be criticized on the same grounds, the one sort of practice can't be invoked as a way of disburdening the other one from the same objection."[122]

---

120. This is also an objection that Elizabeth Fenton raises against Habermas in "Liberal Eugenics and Human Nature", 39: "Habermas's concern that liberal eugenics fundamentally alters the nature of human relationships is overblown. . . . The parent-child relationship is inherently one of inequality; even without explicitly choosing a child's characteristics or traits, a parent has considerable control over the range of options open to her for future development. Moreover, such inequalities or asymmetries in relationships abound within the human moral community."

121. Cf. Junker-Kenny, "Genetic Enhancement as Care or as Domination?" 12: "Adults accompanying children's 'social birth' have to walk a tight line between respect for their difference *and* respect for their dependence. In both, they have to be able to distinguish between their own desires and hopes, and the reality of the other. The biblical command, 'Thou shalt not make an image of me' (cf. Ex 20) captures this attitude. For the Swiss author Max Frisch, making an image of the other person and fixing him to it is 'the ultimate betrayal.' It appears that the foremost pedagogical quality is to be able to control one's own projections."

122. Habermas, *Future of Human Nature*, 84.

Here we should also note that the problem with genetic enhancement, as Habermas sees it, has only little to do with the irreversibility of the intervention, even though Robert Song must have understood the argument to turn on this point. Song draws attention to the fact that in educating their children, parents make irreversible decisions for their children all the time and concludes from this observation that Habermas' case is wanting: "To say that *all* parentally-chosen irreversible changes are morally unacceptable, which consistency requires Habermas to say, is doubtful."[123] However, Habermas nowhere claims that the procedure is questionable *because* it is irreversible.[124] What seems to be much more at stake for him is *where, when,* and *how* the irreversible intervention takes place. It occurs *in the body* at the *moment of its constitution,* depriving the children of their contingent beginning and manipulating their very identity; and it takes place *in an instrumental fashion,* disregarding the children as second persons and treating them as third persons. We can in fact think of many ways in which parents make irreversible decisions for their children by which however they do not instrumentalize them at all but always treat them as second persons. Habermas' concern is not just about any irreversible procedure, but about one that strikes at the core of who the other is and for which there cannot be any presumed consent.

A further objection against Habermas' general argument was raised by Ronald Dworkin who pointed out that the case rests on the assumption that genetically programmed persons will at some point learn about the intervention.[125] Hence, one could forestall all the negative effects on their self-understanding as autonomous agents by simply not telling them about the intervention. Here Habermas responds that the question regarding information about genetic manipulation is similar to information about one's parents. It is a question that touches the origin of persons and thus their very identity. He concludes, to our mind convincingly, that trying to help people in coping with their crisis of identity by concealing

123. Song, "Knowing There Is No God," 200.

124. As we will see shortly, it is true that Habermas would not think the procedure problematic if it were reversible. But from this it does not follow that its irreversibility is itself the reason for why Habermas objects to it. He simply thinks—and to us it seems wrongly so—that by being reversible and thus under the control of the patient's free will, a genetic intervention would cease to be instrumentalizing and could always be considered to take place within a communicative context of second persons.

125. Cf. Habermas, *Future of Human Nature,* 86–87. Habermas refers to Ronald Dworkin as the author of this objection, however, he does not indicate any written source.

from them relevant information about their identity can hardly be called a plausible procedure.[126]

There is a final objection-and-response pair which we would like to mention here, one that is very revealing of Habermas' approach, even though his response could perhaps be challenged even on his own terms. It is again Ronald Dworkin who calls attention to the fact that Habermas' argument tacitly assumes that "the genetic intervention is carried out by a second person, and not by the affected person herself."[127] Implied in this observation is the proposal of a hypothetical scenario in which parents perform enhancing genetic interventions on their children in accordance with their best intentions and at the same time provide for a kind of genetic "switch" by which the adolescents—having come to the use of their reason and in possession of their own vision for life—can later painlessly choose to accept or reject the modification. Habermas replies that he does not see any problem with such a procedure, since what matters for him is the autonomy of persons and their ability to choose and not any supposed inviolability of human nature as it is given, which for him does not deserve any particular respect.[128]

However, we may wonder whether this answer does not dodge the issue, at least the moment we take things to the second level: the capacity to choose a given trait is itself a trait, which, once it is there, is no longer negotiable.[129] The adolescents will *have to choose*. Genetically modifying their children in the way described, the parents do not ask them whether they want to have that option, nor can they presume their consent here. The capacity to choose a trait burdens the children with the responsibility for it in either way, whether they accept or reject it. Their size, hair color,

126. Cf. ibid., 87: "It is hardly permitted to forestall the identity crises of a young person in such a way that one takes the precaution of concealing from him precisely the causal history of the anticipated problem, thereby adding to the programming itself the deception over this relevant fact of life."

127. Ibid., 86.

128. Cf. ibid., 87: "The argument doesn't proceed on the assumption that the technicization of 'inner nature' constitutes something like a transgression of natural boundaries. The criticism remains valid quite independently of the idea of a 'natural' or even 'holy' order, which can be sacrilegiously 'overstepped.' Instead the argument against alien determination draws its strength completely from the fact that a genetic designer, acting according to his own preferences, assumes an irrevocable role in determining the contours of the life history and identity of another person, while remaining unable to assume even her counterfactual consent."

129. I would like to thank Rev. José Granados, DCJM for having alerted me to this problem with Habermas' response.

intelligence, or whatever else one may hypothetically modify genetically, will no longer be a naturally-grown quality, but rather one for which now human persons have to take responsibility: in this case, they themselves. Having to make this choice puts them into an indirect relationship to themselves, insofar as it forces them to act on their body as if it were an object and not part of themselves as subjects. In any case, they may have preferred to be naturally-grown and not having to be responsible for their size etc. And insofar as their parents are responsible for the fact that now they are responsible for this trait, the instrumental relationship between them and their parents continues to be in place just like it was with other forms of genetic manipulation.

Concluding Remarks

In this chapter we have asked about the lasting influence of Hans Jonas for bioethics and for this reason have looked at his place in a more or less recent contribution to the debate by Jürgen Habermas. We found that Habermas has taken up clues from Jonas at central points of his argument. Habermas' central concern about genetic enhancement is that it leads to the domination of some over others, disturbing the fundamental equality that for him has to be present in human relationships if they are to be governed by morality. Even though Jonas does not build his own ethical theory on equality and reciprocity but tries to give an account of responsibility in the face of people's fundamental asymmetry of power, his great concern is about human persons' domination of their own nature, which may lead to intolerable relations of domination of the few over the many, and which may endanger the "image of man" as we know it.[130] Insofar as Jonas points to the danger of unjust domination, Habermas explicitly draws on him in support of his own case. Further, we have seen that Habermas implicitly bases himself on Jonas when he discusses the threat that genetic engineering poses to the spontaneity of human persons and their sense of ownership over their own life and action. And here the two are in plain agreement. The reason for the moral

130. We have argued above that with the certain "inequality" or "nonreciprocity" implied in his concept of responsibility Jonas mainly has in mind asymmetry of power. We have also seen that inequality in the sense of an arbitrary domination of some people over others—in particular of the present generation over the future ones—is a reason for concern for Jonas as much as it is for Habermas. In fact, as we have mentioned, Habermas cites Jonas here in support of his views.

misgivings concerning genetic manipulation are ultimately the same as those regarding cloning: it is the power of an alien intention and expectation engraved in the person's very body for which one can take no recourse. The parents are imposing their own image on their child. Thus the conclusion of Habermas' booklet can be seen to be very similar to the point with which Jonas ends his *The Imperative of Responsibility*: We must preserve the "image of man," which primarily takes on the negative duty to refrain from imposing our own image on him, because "genuine man is always already there and was there throughout known history: in his heights and his depths, his greatness and wretchedness . . . in all the *ambiguity* that is inseparable from humanity. Wishing to abolish this constitutive ambiguity is wishing to abolish man in his unfathomable freedom."[131] At another place Jonas puts this insight into explicitly religious language: "We simply must not try to fixate man in any image of our own definition and thereby cut off the as yet unrevealed promises of the image of God."[132] Or, in other words, as the Old Book says, "You shall not make an image" (cf. Exod 20:4).[133]

131. Jonas, *Imperative of Responsibility*, 200–201.

132. Jonas, *Philosophical Essays*, 181.

133. Cf. Junker-Kenny, "Genetic Enhancement as Care or as Domination?" 12.

Conclusion

A Greater Freedom

WE STARTED OUR REFLECTIONS with a quotation by Francis Fukuyama, in which he expresses his fear that biotechnology may cause us to lose our humanity and our moral sense, which for him is grounded in our human nature: "The deepest fear that people express about technology is . . . that, in the end, biotechnology will cause us in some way to lose our humanity . . . . Human nature is what gives us a moral sense, provides us with the social skills to live in society, and serves as a ground for more sophisticated philosophical discussions of rights, justice, and morality. What is ultimately at stake with biotechnology is . . . the very grounding of the human moral sense, which has been a constant ever since there were human beings."[1] We hope that at the end of this itinerary it may be clearer in which sense this concern is very real. Critics of Fukuyama have challenged him on the grounds that he takes human nature to be a normative point of reference, while supposedly ever since David Hume we know that from a statement about what is the case we cannot derive statements about what ought to be done.[2] There is no bridge from the *is*

1. Fukuyama, *Our Posthuman Future*, 101–2.
2. Cf. Agar, *Liberal Eugenics*, 90–91: "Moral philosophers have long been suspicious of appeals to nature. It is the received view that facts about nature cannot, by themselves, lead to moral conclusions. The logical gap between 'is' and 'ought' means that we can describe how humans typically are, or have been designed to be, without justifying any claim about how they should be. Merely pointing out that humans often lie when they have a strong personal incentive to cannot justify lying without some additional moral claim. Fukuyama needs to cross this gap if he is to show that merely remaining human is some manner of moral achievement."

to the *ought*. A simple description of human nature, on this account, will hence not be sufficient for guiding us in biotechnological questions.

The issue of course turns around the question of what we mean by human nature, or any nature for that matter. The objection of Hume and his followers would seem true if with nature we referred to simple facticity—nature is simply what is the case. Evidently, the fact that someone is "naturally" short-sighted does not ground any norm against wearing glasses that help him or her to see well. The fact that the human being is naturally unable to fly does not serve as grounds for a norm prohibiting the construction of airplanes. We hope that particularly in the first chapter we have seen that an understanding of "nature" as mechanic facticity is inadequate. Nature is never mere facticity; rather it is a dynamic and teleological principle. Already by looking at nature from a biological point of view—in discussing Jonas' understanding of the organism in the first chapter—we have found the intimations of freedom and of love. But human nature in fact goes beyond its organic aspects. The capacities for responsibility and spontaneity, discussed in the second and third chapters, are part and parcel of it. At the same time we must beware of distinguishing too neatly the biological and personal dimensions of human nature, since these are strictly intertwined.

As we have seen in the first chapter, human relationality is indeed based on biology, on the fact, that is, that humans are organisms and as such needful beings—marked by metabolism—and fruitful beings—marked by reproduction. In humans, these biological facts take on a personal significance. Let us just think of the lines of descent. These are certainly biological but go beyond that, establishing filial and parental relations that will forever define who one is. Knowing one's line of descent, knowing "who is whose," is of utmost importance for one's identity.[3] The Hebrew Scriptures are famous for their genealogies, and also the New Testament does not hesitate to trace Jesus' human line of descent down through all the ages back to Adam. It is a fact that still today people are fascinated by royal families. Even after the abolition of the monarchies in Germany, Austria, and Italy, it makes a significant difference whether someone is called von Hohenzollern, von Habsburg, or di Savoia—or Müller, Schuster, or Neri. Even in the United States, which never had a monarchy, the Kennedy family has achieved the fame and admiration of a quasi-royalty. But the issue is not just about aristocracy or royalty. For

---

3. Cf. Kass, *Life, Liberty*, 99–100.

all human beings it seems naturally important to know who their parents are and to have some idea about their line of descent. It is true that, as to what can be measured and scientifically verified about individual persons—their physical and psychological health or lack thereof, their abilities and weaknesses—all these are quite independent of their origin and line of descent. Someone may be the grandson of a great criminal and still be a morally descent person. But does the line of descent really not matter at all? Are we simply superstitious when we feel a certain respect when we are introduced to person called von Habsburg or when we are reluctant to send our children to a school in which a close relative of Adolph Hitler is one of the instructors? No, it is not just superstition. Human identity is constituted by one's origin and destiny. To know who I am, I need to know where I am coming from and where I am going to. My origin is thus a constitutive part of my identity as a human being. People, conceived by artificial insemination or in-vitro fertilization, who are looking for the anonymous donor of the semen by which they came to be, are doing something deeply human. There is a deep human desire and authentic need to know who one's biological parents are.

Human nature is relational and inasmuch as our new biotechnologies touch on this aspect of our existence, we must ask what they do to it. Our relationality is thus going to be a standard for evaluating biotechnological procedures: do they leave the lines of descent intact or do they confuse them, causing confusion also for the person's identity? Besides, a related point is that the human *species* is also the human *family*. In fact, according to the findings of genetics, all our lines of descent meet at some point in the distant past.[4] Hence, another question to ask in the evalua-

---

4. We are referring here to what is called the "Mitochondrial Eve." Cf. Oppenheimer, *Real Eve*, 37: "To say that we get exactly half of our DNA from our father and half from our mother is not quite true. One tiny piece of our DNA is inherited only down the female line. It is called mitochondrial DNA because it is held as a unique circular stand in small tubular packets known as mitochondira that function rather like batteries within the cell cytoplasm. . . . Males, although they receive and use their mother's mitochondrial DNA, cannot pass it on to their children. The sperm has its own mitochondria to power the long journey from the vagina to the ovum but, on entry into the ovum, the male mitochondria wither and die. . . . So each of us inherits our mtDNA from our own mother, who inherited her mtDNA intact from her mother, and so on back through the generations—hence mtDNA's popular name, 'the Eve gene.' Ultimately, every person alive today has inherited their mitochondrial DNA from one single great-great-great- . . . - grandmother, nearly 200,000 years ago." No claims are made by scientists that the Mitochondrial Eve is the first woman ever to have lived, only that all humans alive today can trace their decent to her, which is enough to

tion of biotechnology is what given procedures do to the unity of this family.

Then, as we have seen in chapter two, it is part of human nature to be responsible. Human beings are naturally responsible beings, capable of appreciating and caring for what is precious and precarious. Benevolence is the joy in the happiness of others, the willing of the others' good, when it is in our power to do it and particularly when they are in need. This, we have argued, is the natural attitude of a rational being. What needs justification is not why someone should be benevolent or responsible in the face of a precious being in need, but rather why he or she should not be. Thus, taking human nature as a standard for judging the procedures of biotechnology, we must ask ourselves: what do these technologies do to our capacity to allow others to become real to us? Might they manipulate our brains to the extent of causing us to remain locked in ourselves, unable to transcend ourselves and to perceive the other as another self? The central question, after all, is not only whether they succeed in making us more cheerful, more intelligent, and stronger—all of which are attributes compatible with living in Huxley's *Brave New World*—but rather what they do to our responsibility. *Humanly* speaking, a person who is often sad, who is not the brightest bulb in the chandelier, and who is physically rather weak, yet capable of benevolence or responsibility is much more to be preferred than a cheerful athletic genius who has lost his or her capacity to love.

Finally, in chapter three we have seen that it is part of human nature to be spontaneous and to take initiative. In fact, the Latin word *natura* also means birth, and it is in the fact that human beings are born, in their "natality," that Hannah Arendt sees the ontological root of their capacity to begin something new.[5] Thus, nature and spontaneity must not be opposed as if nature were something static and spontaneity and freedom consisted in modeling and shaping nature as one pleases. Rather, nature itself is a principle of spontaneity, a source of action and novelty. Taking nature as a criterion for the moral evaluation of biotechnology means,

---

affirm the unity of the human species. Cf. Dawkins, *River Out of Eden*, 54: "The correct claim is not that she [the Mitochondrial Eve] was the only woman on Earth, nor even that the population was relatively small during her time. Her companions, of both sexes, may have been both numerous and fecund. . . . The correct claim is only that Mitochondrial Eve is the most recent woman of whom it can be said that all modern humans are descended from her in the female-only line."

5. Cf. Arendt, *Human Condition*, 247.

asking with Jürgen Habermas, whether a given procedure might not undermine our capacity to see ourselves as the undivided authors of our own lives, to find within *ourselves* the source of our action and reaction or whether this technique forces us to share the authorship of our lives with others, thus obstructing our spontaneity and impeding our freedom. Does a given technology, with all the good intentions with which it is proposed, still allow us to call our actions—and ultimately our lives, which are the sum of our actions—really our own, or does it modify us in a way that alienates us from ourselves? Does it perhaps hinder us from being the protagonists of our own activity, either because it enshrines alien intentions into the constitution of our very bodies or because it substitutes our own efforts and will power for a technological procedure? Are the wonderful gifts and talents we give to ourselves by means of our tools still our own? To be free means to be the principle of one's own activity. Where biotechnology threatens to replace our own authorship of our lives, there it does not lead to a greater freedom but rather serves to diminish it.

But when exactly does biotechnology threaten to stifle our initiative, alienate us from our own activity, cause us to become irresponsible and confuse our relational identity? The answers will at times be rather clear, and at other times they will be rather complex and difficult. The President's Council makes a convincing case that doping alienates athletes from the excellence of their performance[6]; Habermas advances persuasive arguments that a future practice of genetic engineering could alienate persons thus engineered from their body[7]; Jonas compellingly warns us against a world in which, with the alchemy of our biotechnology, we manage to create people content but incapable of responsibility[8], and Kass forcefully shows how artificial procreation can do much harm to confuse people's lines of descent, making it hard for them to find their identity[9]. These concerns however, are always about the discernment of specific procedures. The question is certainly not "Biotechnology: yes or no?" There is no doubt that biotechnology has a already achieved much good and that it has a great future promise for yet more good. It does promise a greater freedom that comes with increased strength, more

6. Cf. The President's Council on Bioethics, *Beyond Therapy*, 101–57.

7. Cf. Habermas, *Future of Human Nature*, 57–60.

8. Cf. Jonas, *Imperative of Responsibility*, 42.

9. Cf. Kass, *Life, Liberty*, 81–117.

talents, greater intelligence, and longer days. And who could be against this?

At the same time, we must not forget that the aspects of our lives which are open to technological enhancement are not all there is to us. A problem comes in the moment biotechnological engineering makes claim to a comprehensive worldview. At this point, then, we are faced with the alternative once formulated by Joseph Ratzinger in a reflection on artificial procreation: "One can on the one hand regard only the mechanical, nature's laws, as real, and consider all that is personal, love, giving, as pretty appearance, which, though psychologically useful, is ultimately unreal and untenable. . . . Next to this, according to the other alternative, things are just the opposite: One can consider the personal as the real, the stronger and higher form of reality, which does not reduce the other realities, the biological and mechanical, to mere appearance, but absorbs them into itself, and thus opens to them a new dimension."[10] For the former position he can find "no other designation than the denial of humanity."[11] In this essay we have sought to argue in favor of the latter stance, i.e., for the centrality of the personal dimension. The use of our biotechnologies, with all their promises of enhancing the range of possibilities for our freedom, must thus never be to the detriment of the sphere of interpersonal relationships. Indeed, the greater freedom is the freedom for our destiny: the freedom to be the responsible, benevolent, spontaneous authors of our own lives, capable of entering into relationships with others, living lives of communion and love. The touchstone for our evaluation of any proposed biotechnological procedure has to be this greater freedom.

10. Ratzinger, "Man between Reproduction and Creation," 209–10.
11. Ibid., 209.

Bibliography

Adams, Nicholas. *Habermas and Theology.* Cambridge: Cambridge University Press, 2006.

Agar, Nicholas. "Liberal Eugenics." In *Bioethics: An Anthology,* edited by Helga Kuhse and Peter Singer, 171–81. Oxford: Blackwell, 1999.

———. *Liberal Eugenics. In Defence of Human Enhancement.* Oxford: Blackwell Publishing, 2004.

Apel, Karl-Otto. "The Ecological Crisis as a Problem for Discourse Ethics." In *Ecology and Ethics: A Report from the Melbu Conference,* edited by Audun Øfsti, 219–60. Trondheim, Norway: Nordland Akademi for Kunst og Vitenskap, 1992.

———. "The Problem of a Macroethic of Responsibility to the Future in the Crisis of Technological Civilization: An Attempt to Come to Terms with Hans Jonas's 'Principle of Responsibility.'" *Man and World* 20 (1987) 3–40.

Aquinas, Thomas. *Commentary on Aristotle's De anima.* Translated by Kenelm Foster and Silvester Humphries. South Bend, IN: Dumb Ox, 1994.

———. *Commentary on Aristotle's Physics.* Translated by Richard J. Blackwell et al. South Bend, IN: Dumb Ox, 1999.

———. *Summa Contra Gentiles, Book III, Part 1 and 2.* Translated by Vernon J. Bourke. Notre Dame, IN: University of Notre Dame Press, 1991.

———. *Summa Theologica.* Translated by The Fathers of the English Dominican Province. Westminster, MD: Christian Classics, 1981.

Arendt, Hannah. "The Eggs Speak Up." In *Essays in Understanding, 1930–1954.* Edited by Jerome Kohn, 270–84. New York: Harcourt, 1994.

———. *The Human Condition.* 2nd ed. Chicago: The University of Chicago Press, 1998.

———. *The Life of the Mind.* Edited by Mary McCarthy. New York: Harcourt, 1978.

Aristotle. *Metaphysics.* Translated by W. D. Ross. Oxford: Oxford University Press, 1924.

———. *Nicomachean Ethics.* Translated by W. D. Ross. Oxford: Clarendon Press, 1908.

———. *Parts of Animals. Movement of Animals. Progression of Animals.* Translated by A. L. Peck and E. S. Forster. Loeb Classical Library. Cambridge: Harvard University Press, 1937.

———. *Physics.* Translated by R. P. Hardie and R. K. Gaye. Whitefish, MT: Kessinger, 2004.

———. *Politics.* Translated by Harris Rackham. Loeb Classical Library. Cambridge: Harvard University Press, 1932.

―――. *On the Soul*. Translated by J. A. Smith. In *The Basic Works of Aristotle*. Edited by Richard McKeon. New York: Random House, 1941.

Augustine. *Confessions*. Translated by John K. Ryan. New York: Doubleday, 1960.

Bacon, Francis. *Of the Dignity and Advancement of Learning*. In *The Works of Francis Bacon*, edited by James Spedding et al., vol. VIII. Boston: Taggard & Thompson, 1863.

―――. *Meditationes sacrae*. In *The Works of Francis Baconm*, edited by James Spedding et al. Vol. XIV. Boston: Taggard and Thompson, 1864.

―――. *The New Organon*. Edited by L. Jardine and M. Silverthorne. Cambridge, UK: Cambridge University Press, 2000.

Bausch, Thomas, et al. *Wirtschaft und Ethik. Strategien contra Moral?* Münster: LIT, 2004.

Bayertz, Kurt. "Human Nature: How Normative Might it Be?" *Journal of Medicine and Philosophy* 28 (2003) 131–50.

Benedict XVI. *Address during the Meeting with the Authorities and the Diplomatic Corps*, Hofburg, Vienna, September 7, 2007.

―――. Encyclical Letter *Deus Caritas Est*, December 25, 2006.

Bernstein, Richard J. "Rethinking Responsibility." *Hastings Center Report* 25/7, Special Issue (1995) 13–20.

Birnbacher, Dieter. "Habermas' ehrgeiziges Beweisziel—erreicht oder verfehlt?" *Deutsche Zeitschrift für Philosophie* 50 (2002) 121–26.

Bloch, Ernst. *The Principle of Hope*. Translated by Neville Plaice et al. Cambridge: MIT Press, 1986.

Blondel, Maurice. *Action (1893). Essay on a Critique of Life and a Science of Practice*. Translated by Oliva Blanchette. Notre Dame, IN: University of Notre Dame Press, 2003.

Böhler, Dietrich, editor. *Ethik für die Zukunft. Im Diskurs mit Hans Jonas*. Munich: C. H. Beck, 1994.

―――. "In dubio contra projectum. Mensch und Natur im Spannungsfeld von Verstehen, Konstruieren und Verantworten." In *Ethik für die Zukunft. Im Diskurs mit Hans Jonas,* edited by Dietrich Böhler, 244–76. Munich: C. H. Beck: 1994.

―――. "Zukunftsverantwortung, Moralprinzip und kommunikative Diskurse. Die Berliner Auseinandersetzung mit Hans Jonas: Grundlagen der Moral, der Wirtschafts- und Bioethik." In *Wirtschaft und Ethik. Strategien contra Moral?* edited by Thomas Bausch et al., 215–88. Münster: LIT, 2004.

Bostrom, Nick. "In Defense of Posthuman Dignity." *Bioethics* 19 (2005) 202–14.

Botturi, Francesco. *Tempo, linguaggio e azione. Le strutture vichiane della "storia ideale eterna."* Naples: Guida, 1996.

Brecht, Bertholt. "The Buddha's Parable of the Burning House." In *Bertholt Brecht: Poems 1913–56*, edited by John Willett and Ralph Manheim, 290–92. New York: Routledge, 1987.

Buchanan, Allen, et al. *From Chance to Choice: Genetics and Justice*. Cambridge, UK: Cambridge University Press, 2000.

Camus, Albert. *The Myth of Sisyphus and Other Essay*. Translated by Justin O'Brian. New York: Vintage, 1991.

Cohen, Adam. "The Unalienable Rights of Chimps?" *International Herald Tribune* (July 16, 2008) 11.

Colombo, Roberto. "Vita: dalla biologia all'etica." In *Quale vita? La bioetica in questione*, edited by Angelo Scola, 169–95. Milan: Mondadori, 1998.

Congregation for the Doctrine of the Faith. *Donum Vitae. Instruction on Respect for Human Life in Its Origin and on the Dignity of Human Procreation*. February 22, 1987.

Descartes, René. *The Correspondence*. In *The Philosophical Writings of Descartes*, vol. III. Translated by John Cottingham et al. Cambridge, UK: Cambridge University Press, 1991.

———. *Meditations on First Philosophy*. Translated by J. Cottingham. Cambridge, UK: Cambridge University Press, 1996.

———. *The Passions of the Soul*. Translated by Stephen Voss. Indianapolis, IN: Hackett, 1989.

Dini, Alessandro. "Natura umana e biotecnologia. Recenti interventi e quadro storico." *Annali del Dipartimento di Filosofia* 11 (2006) 33–56.

Dumitru Nalin, Speranta. "Liberté de procréation et manipulation génétique. Pour une critique d'Habermas." *Raisons Politiques* 12 (2003) 31–54.

Dawkins, Richard. *River Out of Eden. A Darwinian View of Life*. New York: Basic, 1996.

Dworkin, Ronald. *Sovereign Virtue. The Theory and Practice of Equality*. Cambridge: Harvard University Press, 2002.

Ebner, Ferdinand. *Das Wort und die geistlichen Realitäten*. Frankfurt/Main: Suhrkamp, 1980.

Edgar, Andrew. *The Philosophy of Habermas*. Chesham Buck, UK: Acumen, 2005.

Elliott, Carl. *Better than Well. American Medicine Meets the American Dream*. New York: W. W. Norton, 2003.

Fenton, Elizabeth. "Liberal Eugenics and Human Nature: Against Habermas." *Hastings Center Report* 36 (2006) 35–42.

Finlayson, James Gordon. *Habermas. A Very Short Introduction*. Oxford: Oxford University Press, 2005.

Frankena, William K. *Ethics*, Englewood Cliffs, NJ: Prentice-Hall, 1963.

———. *Ethics*. 2nd ed. Englewood Cliffs, NJ: Prentice-Hall, 1973.

Fukuyama, Francis. *Our Posthuman Future: Consequences of the Biotechnology Revolution*. New York: Picador, 2002.

Garreau, Joel. *Radical Evolution: The Promise and Peril of Enhancing Our Minds, Our Bodies—and What It Means to Be Human*. New York: Broadway, 2005.

Gensabella Furnari, Marianna. "From the Ontology of Temporality to the Ethics of the Future." In *Philosophy and Ethics: New Research*, edited by Laura V. Siegal, 131–55. Commack, NY: Nova Science, 2006.

Gill, Mary Louise. "Matter Against Substance." *Synthese* 96 (1993) 379–97.

Granados, José. "Love and the Organism: A Theological Contribution to the Study of Life." *Communio* 32 (2005) 435–69.

Habermas, Jürgen. "Auf schiefer Ebene." *Die Zeit* (January 24, 2002) 33.

———. *Erläuterungen zur Diskursethik*. Frankfurt/Main: Suhrkamp, 1991.

———. *The Future of Human Nature*. Cambridge, UK: Polity Press, 2003.

———. *Justification and Application: Remarks on Discourse Ethics*. Translated by Ciaran P. Cronin. Cambridge: MIT Press, 1994.

———. "Modernity: An Unfinished Project." In *Habermas and the Unfinished Project of Modernity*, edited by Maurizio Passerin d'Entrèves and Seyla Benhabib, 38–55. Cambridge: MIT Press, 1997.

——. *Moral Consciousness and Communicative Action.* Translated by Christian Lenhardt and Shierry Weber Nicholsen. Cambridge: MIT Press: 1990.

——. *Postmetaphysical Thinking. Philosophical Essays.* Translated by William Mark Hohengarten. Cambridge: MIT Press, 1994.

——. *The Structural Transformation of the Public Sphere: An Inquiry into a Category of Bourgeois Society.* Translated by T. Burger and F. Lawrence. Cambridge: MIT Press, 1989.

——. *The Theory of Communicative Action,* vol. 1: *Reason and the Rationalization of Society.* Translated by Thomas McCarthy. Boston: Beacon, 1984.

——. *The Theory of Communicative Action,* vol. 2: *Lifeworld and System: A Critique of Functionalist Reason.* Translated by Thomas McCarthy. Boston: Beacon, 1987.

——. *Zeit der Übergänge.* Frankfurt/Main: Suhrkamp 2001.

——. *Die Zukunft der menschlichen Natur. Auf dem Weg zu einer liberalen Eugenik?* Frankfurt/Main: Suhrkamp, 2001.

Habermas, Jürgen and Joseph Ratzinger. *The Dialectics of Secularization. On Reason and Religion.* San Francisco: Ignatius, 2006.

Hauerwas, Stanley. *Dispatches from the Front. Theological Engagements with the Secular.* Durham, NC: Duke University Press, 1994.

——. *Suffering Presence: Theological Reflections on Medicine, the Mentally Handicapped, and the Church.* Notre Dame, IN: University of Notre Dame Press, 1986.

——. *Truthfulness and Tragedy.* Notre Dame, IN: University of Notre Dame Press, 1977.

Hauerwas, Stanley, and Charles Pinches. *Christians Among the Virtues: Theological Conversations with Ancient and Modern Ethics.* Notre Dame, IN: University of Notre Dame Press, 1997.

Heidegger, Martin. *Being and Time.* Translated by John Macquarrie and Eward Robinson. New York: Harper & Row, 1962.

Hobbes, Thomas. *Leviathan.* Edited by J. Gaskin. Oxford: Oxford University Press, 1996.

Hofheinz, Marco. *Gezeugt, nicht gemacht.* Münster: LIT, 2008.

Horkheimer, Max and Theodor Adorno. *Dialectic of Enlightenment.* Translated by Edmund Jephcott. Palo Alto: Stanford University Press, 2007.

Hösle, Vittorio. "Ontologie und Ethik bei Hans Jonas." In *Ethik für die Zukunft. Im Diskurs mit Hans Jonas,* edited by Dietrich Böhler, 105–25. Munich: C. H. Beck, 1994.

Hottois, Gilbert. "Un'analisi critica del neo-finalismo della filosofia di H. Jonas." *Idee* 26/27 (1994) 85–105.

Hume, David. *An Enquiry Concerning Human Understanding.* Edited by Stephen Buckle. Cambridge, UK: Cambridge University Press, 2007.

Huxley, Aldous. *Brave New World.* London: Chatto and Windus, 1932.

James, William "The Moral Philosopher and the Moral Life." In *Pragmatism and Other Writing,* edited by Giles B. Dunn, 242–64. New York: Penguin, 2000.

John Paul II. Encyclical Letter *Evangelium vitae,* March 25, 1995.

——. Encyclical Letter *Veritatis Splendor,* August 6, 1993.

Jonas, Hans. *Augustin und das paulinische Freiheitsproblem: Eine Beitrag zur Entstehung des christlich-abendländischen Freiheitsbegriffs.* Göttingen: Vandenhoeck & Ruprecht, 1930.

———. "Biological Foundations of Individuality." *International Philosophical Quarterly* 8 (1968) 231–51.

———. "The Burden and Blessing of Mortality." *Hastings Center Report* 22 (1992) 34–40.

———. "Change and Permanence: On the Possibility of Understanding History." *Social Research* 38 (1971) 498–528.

———. "Ethics and Biogenetic Art." *Social Research* 71 (2004) 569–82.

———. "Fatalismus wäre Todsünde." In *Ethik für die Zukunft. Im Diskurs mit Hans Jonas*, edited by Dietrich Böhler, 455–56. Munich: C. H. Beck, 1994.

———. *Gnosis und spätantiker Geist*. Göttingen: Vandenhoeck & Ruprecht, 1934.

———. *The Gnostic Religion: The Message of the Alien God and the Beginnings of Christianity*. Boston: Beacon, 1958.

———. "Gnosticism and Modern Nihilism." *Social Research* 19 (1952) 430–52.

———. *The Imperative of Responsibility: In Search of an Ethics for the Technological Age*. Chicago: University of Chicago Press, 1984.

———. *Memoirs: Hans Jonas*. Edited by Christian Wiese and translated by Krishna Winston. Waltham, Mass.: Brandeis, 2008.

———. *Mortality and Morality: A Search for the Good after Auschwitz*. Edited by Lawrence Vogel. Northwestern University Press, Evanston, Ill., 1996.

———. *The Phenomenon of Life: Toward a Philosophical Biology*. Evanston, Ill.: Northwestern University Press, 2001.

———. *Philosophical Essays: From Ancient Creed to Technological Man*. Chicago: University of Chicago Press, 1974.

———. "Philosophy at the End of the Century: A Survey of Its Past and Future." *Social Research* 61 (1994) 813–32.

———. *Das Prinzip Verantwortung*. Frankfurt/Main: Suhrkamp, 1979.

———. *Technik, Medizin und Ethik. Zur Praxis des Prinzips Verantwortung*. Suhrkamp: Frankfurt/Main, 1987.

———. "Toward a Philosophy of Technology." *Hastings Center Report* 9 (1979) 34–43.

Jonas, Hans and Harvey Scodel. "An Interview with Professor Hans Jonas." *Social Research* 70 (2003) 339–68.

Junker-Kenny, Maureen. "Genetic Enhancement as Care or as Domination? The Ethics of Asymmetrical Relationships in the Upbringing of Children." *Journal of Philosophy of Education* 39 (2005) 1–17.

Kaminski, Juliane, et al. "Word Learning in a Domestic Dog: Evidence for 'Fast Mapping.'" *Science* 304 (June 11, 2004) 1682–83.

Kant, Immanuel. *Groundwork of the Metaphysics of Morals*. Translated by H. J. Paton. New York: Harper Torchbooks, 1964.

Kass, Leon R. "Appreciating *The Phenomenon of Life*." *Hastings Center Report* 25/7, Special Issue (1995) 3–12.

———. *Life, Liberty and the Defense of Dignity. The Challenge for Bioethics*. San Francisco: Encounter, 2002.

———. "Practicing Ethics: Where's the Action?" *Hastings Center Report* 20 (1990) 5–12.

———. *Toward a More Natural Science. Biology and Human Affairs*. New York: The Free Press, 1985.

Kuhlmann, Andreas. "Wider die Verdinglichung des Menschen." *Die Zeit* 39 (September 20, 2001) 55.

Bibliography

Kuhlmann, Wolfgang. "'Prinzip Verantwortung' versus Diskursethik." In *Ethik für die Zukunft. Im Diskurs mit Hans Jonas,* edited by Dietrich Böhler, 277–302. Munich: C. H. Beck, 1994.

Leibniz, Gottfried Wilhelm. "Codex Iuris Gentium (Praefatio) (1693)." In *Political Writings,* edited by Patrick Riley. Cambridge: Cambridge University Press, 1988.

Lewis, Clive Staples. *The Abolition of Man.* New York: Macmillan, 1947.

Löw, Reinhard. "Zur Wiederbegründung der organischen Naturphilosophie durch Hans Jonas." In *Ethik für die Zukunft. Im Diskurs mit Hans Jonas,* edited by Dietrich Böhler, 68–79. Munich: C. H. Beck, 1994.

Madison. Gary B. "Critical Theory and Hermeneutics. Some Outstanding Issues on the Debate." In *Perspectives on Habermas,* edited by Lewis E. Hahn, 463–85. Chicago: Open Court, 2000.

Malmqvist, Erik. "Back to the Future: Habermas's *The Future of Human Nature.*" *Hastings Center Report* 37 (March–April 2007) 5.

Marcel, Gabriel. *Homo Viator: Introduction to a Metaphysics of Hope.* Translated by E. Crawford. New York: Harper & Row, 1962.

Marx, Karl. *Capital: A New Abridgment.* Edited by David McLellan. Oxford: Oxford University Press, 1995.

———. *Critique of the Gotha Programme.* In *Selected Writings.* Edited by Lawrence H. Simon. Indianapolis, IN: Hackett, 1994.

McKenny, Gerald P. *To Relieve the Human Condition.* Albany: State University of New York Press, 1997.

Melina, Livio. *Building a Culture of the Family: The Language of Love.* Staten Island, NY: The Society of St. Paul—Alba House, 2011.

———. *La conoscenza della morale. Linee di riflessione sul Commento di San Tommaso all'etica Nicomachea.* Rome: Città Nuova, 1987.

———. *The Epiphany of Love. Toward a Theological Understanding of Christian Action.* Grand Rapids: Eerdmans, 2010.

———. "La prudenza secondo Tommaso d'Aquino," *Aquinas* 49 (2006) 381–94.

———. *Sharing in Christ's Virtues. For a Renewal of Moral Theology in Light of Veritatis Splendor.* Translated by William E. May. Washington, DC: The Catholic University of America Press, 2001.

———. "Vita." In *Dizionario interdisciplinare di scienza e fede. Cultura scientifica, filosofia e teologia,* Vol. II, edited by Giuseppe Tanzella-Nitti and Alberto Strumia, 1519–30. Rome: Urbania University Press—Città Nuova, 2002.

Melina, Livio, et al. *Camminare nella luce dell'amore. I fondamenti della morale cristiana.* Siena: Cantagalli, 2008.

Mendieta, Eduardo. "Habermas on Human Cloning. The Debate on the Future of the Species." *Philosophy and Social Criticism* 30 (2004) 721–43.

Moss, Lenny. "Contra Habermas and Towards a Critical Theory of Human Nature and the Question of Genetic Enhancement." *New Formations* 60 (Winter 2006–2007) 139–49.

Naam, Ramez. *More Than Human: Embracing the Promise of Biological Enhancement.* New York: Broadway, 2005.

Nathanson, Neal, et al. "Bovine Spongiform Encephalopathy (BSE) Causes and Consequences of a Common Source Epidemic." *American Journal of Epidemiology* 145 (1997) 959–69.

Neubach, Gangolf. *Das Selbstseinkönnen eingebettet in der Gattungsethik: Das postmetaphysische Moralverständnis in Bezug auf "Die Zukunft der menschlichen Natur" von Jürgen Habermas.* Munich: Grin, 2008.

Nietzsche, Friedrich. *The Genealogy of Morals.* Translated by Douglas Smith. Oxford: Oxford University Press, 1998.

———. *The Will to Power.* Edited by Walter Kaufmann and translated by Walter Kaufmann and R. J. Hollingdale. New York: Vintage, 1968.

O'Donovan, Oliver. *Begotten or Made?* Oxford: Clarendon, 1984.

Oppenheimer, Stephen. *The Real Eve: Modern Man's Journey out of Africa.* New York: Basic, 2004.

Palmer, Richard E. "Habermas versus Gadamer? Some Remarks." In *Perspectives on Habermas,* edited by Lewis E. Hahn, 487–500. Chicago: Open Court, 2000.

Péguy, Charles. *Portal of the Mystery of Hope.* Translated by David L. Schindler, Jr. New York: Continuum, 2005.

Pérez-Soba, Juan-José. "*Operari sequitur esse?*" In Livio Melina et al. *La plenitud del obrar cristiano: dinámica de la acción y perspectiva teológica de la moral,* 65–83. Madrid: Palabra, 2001.

Plato. *Gorgias.* In *Lysis. Symposium. Gorgias.* Translated by W. R. M. Lamb. Cambridge: Harvard University Press, 1925.

Plessner, Helmuth. *Die Stufen des Organischen und der Mensch.* Berlin: de Gruyter, 1928.

Polanyi, Michael. "Life's Irreducible Structure." *Science* 160 (1968) 1308–12.

Portinaro, Pier Paolo. "Il profeta e il tiranno. Considerazioni sulla proposta filosofica di Hans Jonas." *Nuova civiltà delle macchine* 37 (1992) 100–111.

The President's Council on Bioethics. *Beyond Therapy. Biotechnology and the Pursuit of Happiness.* New York: Regan, 2003.

Prusak, Bernard G. "Rethinking 'Liberal Eugenics': Reflections and Questions on Habermas on Bioethics." *Hastings Center Report* 35 (2005) 31–42.

Pyle, Andrew. *Malebranche.* New York: Routledge, 2003.

Rahner, Karl. "Zum Problem der genetischen Manipulation." In *Schriften zur Theologie,* vol. 8, 286–321. Einsiedeln: Benziger, 1967.

Ratzinger, Joseph. "Europe in the Crisis of Cultures." In *L'Europa nella crisi delle culture,* 33–42. Siena: Cantagalli, 2005.

———. "Man between Reproduction and Creation: Theological Questions on the Origin of Human Life." *Communio* 16 (1989) 197–211.

Rhonheimer, Martin. "The Instrumentalization of Human Life." In *Ethics of Procreation and the Defense of Human Life: Contraception, Artificial Fertilization and Abortion,* 153–78, edited by William F. Murphy Jr. Washington, DC: The Catholic University of America Press, 2010.

Rousseau, Jean-Jacques. "Discourse on the Origin and Foundations of Inequality." In *The First and Second Discourses.* Translated and edited by R. and J. Masters. New York: St. Martin's, 1964.

Rozemond, Marleen. "Descartes on Mind-Body Interaction: What's the Problem?" *Journal of the History of Philosophy* 37 (1999) 435–67.

Russo, Nicola. *La biologia filosofica di Hans Jonas.* Naples: Guida, 2004.

Sansonetti, Giuliano. "Un'etica della responsabilità: Hans Jonas." *Humanitas* 47 (1992) 476–90.

Sartre, Jean-Paul. *Nausea.* Translated by Lloyd Alexander. New York: Penguin, 1965.

Bibliography

Schindler, David L. *Heart of the World, Center of the Church. Communio Ecclesiology, Liberalism, and Liberation.* Grand Rapids: Eerdmans, 1996.

Schmidt, Jan C. "Die Aktualität der Ethik von Hans Jonas. Eine Kritik der Kritik des Prinzips Verantwortung." *Deutsche Zeitschrift für Philosophie* 55 (2007) 545–69.

Scola, Angelo, editor. *Quale vita? La bioetica in questione.* Milan: Mondadori, 1998.

Sève, Bernard. "Hans Jonas et l'éthique de la responsabilité." *Esprit* 165 (1990) 72–88

Sève, Lucien. *Pour une critique de la raison bioéthique.* Paris: Odile Jacob, 1994.

Siep, Ludwig. "Moral und Gattungsethik." *Deutsche Zeitschrift für Philosophie* 50 (2002) 111–20.

Singer, Peter. *Practical Ethics.* Cambridge: Cambridge University Press, 1993.

———. "Shopping at the Genetic Supermarket." In *Bioethics in Asia in the 21st Century,* edited by Song Sang-Yong et al., 143–56. Christchurch, NZ: Eubios Ethics Institute, 2003.

Sokolowski, Robert. "Matter, Elements and Substance in Aristotle." *Journal of the History of Philosophy* 8 (1970) 263–88.

Song, Robert. "Knowing There Is No God, Still We Should Not Play God? Habermas on the Future of Human Nature." *Ecotheology* 11 (2006) 191–211.

Spaemann, Robert. *Basic Moral Concepts.* Translated by Timothy J. Armstrong. London: Routledge, 1989.

———. "Gezeugt, nicht gemacht." *Die Zeit* (January 18, 2001) 37–38.

———. "Habermas über Bioethik." *Deutsche Zeitschrift für Philosophie* 50 (2002) 105–9.

———. *Happiness and Benevolence.* Translated by Jeremiah Alberg. Notre Dame, IN: University of Notre Dame Press, 2000.

———. "Kommentar." In Kongregation für die Glaubenslehre, *Die Unantastbarkeit des menschlichen Lebens—Zu ethischen Fragen der Biomedizin,* Herder: Freiburg i.Br. 1987.

———. *Persons. The Difference between "Someone" and "Something."* Translated by Oliver O'Donovan. Oxford: Oxford University Press, 2006.

———. *Schritte über uns hinaus. Gesammelte Reden und Aufsätze I.* Stuttgart: Klett-Cotta, 2010.

———. "Wer hat wofür Verantwortung? Kritische Überlegungen zur Unterscheidung von Gesinnungsethik und Verantwortungsethik." In *Grenzen. Zur ethischen Dimension des Handelns,* 218–37. Stuttgart: Klett-Cotta, 2001.

Spaemann, Robert and Reinhard Löw. *Natürliche Ziele. Geschichte und Wiederentdeckung des teleologischen Denkens.* Stuttgart: Klett-Cotta, 2005.

Stanley, Wendell M. "Penrose Memorial Lecture. On the Nature of Viruses, Cancer, Genes, and Life—A Declaration of Dependence." *Proceedings of the American Philosophical Society* 101 (August 16, 1957) 317–24.

Stirk, Peter M. R. *Critical Theory, Politics, and Society: An Introduction.* London: Continuum, 2005.

Stock, Gregory. *Redesigning Humans. Our Inevitable Genetic Future.* Boston: Houghton Mifflin Harcourt, 2002.

Svenaeus, Fredrik. "A Heideggerian Defense of Therapeutic Cloning." *Theoretical Medicine and Bioethics* 28 (2007) 31–62.

Viano, Carlo Augusto. "Antiche ragioni per nuove paure: Habermas e la genetica." *Rivista di filosofia* 95 (2004) 277–96.

Vogel, Lawrence. "Foreword." In Hans Jonas, *The Phenomenon of Life. Toward a Philosophical Biology,* xi-xxi. Evanston, IL: Northwestern University Press, 2001.

———. "Hans Jonas's Exodus: From German Existentialism to Post-Holocaust Theology." In Hans Jonas, *Mortality and Morality: A Search for the Good after Auschwitz.* Edited by Lawrence Vogel, 1–40. Evanston, IL: Northwestern University Press, 1996.

———. "Natural Law Judaism? The Genesis of Bioethics in Hans Jonas, Leo Strauss, and Leon Kass." *Hastings Center Report* 36 (2006) 32–44.

Waldstein, Michael. "Hans Jonas' Construct 'Gnosticism.' Analysis and Critique." *Journal of Early Christian Studies* 8 (2000) 341–72.

Wendnagel, Johannes. *Ethische Neubesinnung als Ausweg aus der Weltkrise? Ein Gespräch mit dem 'Prinzip Verantwortung' von Hans Jonas.* Würzburg: Koenigshausen & Neumann, 1990.

Wilson, Michael. *Microbial Inhabitants of Humans. Their Ecology and Role in Health and Disease.* Cambridge: Cambridge University Press, 2004.

Zander, Hans Conrad. *Kurzgefasste Verteidigung der Heiligen Inquisition.* Gütersloh: Gütersloher, 2007.

Zweibel, Ken, et al. "Solar Grand Plan." *Scientific American* 298 (January 2008) 64–73.

www.ingramcontent.com/pod-product-compliance
Lightning Source LLC
Chambersburg PA
CBHW061735270326

41928CB00011B/2239